Marine Biotechnology

ENABLING SOLUTIONS FOR OCEAN
PRODUCTIVITY AND SUSTAINABILITY

OECD
BETTER POLICIES FOR BETTER LIVES

Please cite this publication as:
OECD (2013), *Marine Biotechnology: Enabling Solutions for Ocean Productivity and Sustainability*, OECD Publishing.
http://dx.doi.org/10.1787/9789264194243-en

ISBN 978-92-64-19423-6 (print)
ISBN 978-92-64-19424-3 (PDF)

Photo credits: Cover © 2436digitalavenue – Fotolia.com, © Helen-light/Shutterstock.com .

Corrigenda to OECD publications may be found on line at: *www.oecd.org/publishing/corrigenda*.

Foreword

This report considers how marine biotechnology can contribute to economic and social prosperity and offer potential solutions to global challenges by making use of recent advances in science and technology. It looks at how these advances are improving our understanding of marine life and facilitating access to and study of marine organisms and ecosystems and their largely untapped potential. It examines the challenges associated with the development of these resources, which exist in complex ecosystems and are distributed throughout a vast shared environment. The report argues for a new global framework for the sustainable development of marine biotechnology and identifies areas that may benefit from greater attention as governments develop policies to support marine biotechnology. It also notes some early policy lessons learned by governments that have sought to benefit from bioresources.

An OECD Global Forum on Biotechnology was held in Vancouver, Canada, in 2012 to discuss the opportunities and challenges of marine biotechnology. The forum, entitled *Marine Biotechnology – Enabling Solutions for Ocean Productivity and Sustainability*, brought together policy makers, regulators, industry leaders, academics and social and natural scientists from the 34 OECD countries and from non-member and developing countries to review recent research and debates and to discuss how to realise the potential of marine biotechnology. Insights gained from expert speakers and roundtable discussions over two days were combined with substantive background research by the OECD's Working Party on Biotechnology in order to define opportunities as well as areas in need of further attention. This report presents a synthesis of that work.

The report was drafted by Rachael Ritchie and Jim Philp under the direction of the Working Party on Biotechnology (WPB). Special thanks go to the WPB steering group (composed of representatives from Belgium, Canada, Denmark, Israel, Korea, Mexico, Norway, Switzerland, the United States, the European Union and BIAC); Genome Canada, Genome British Columbia, Genome Atlantic, Health Canada, the Research Council of Norway, the Norwegian Ministry of Trade and Industry, and the Korean Ministry of Land, Transport and Maritime Affairs, which provided support for the Vancouver meeting; and Jody Wright of the University of British Columbia, whose summary of the Vancouver workshop made an invaluable contribution to this report.

Table of contents

Figures

Tables

Executive summary

Interest in marine biotechnology has grown as a result of scientific advances that have increased our knowledge of marine biodiversity and the development of technology and tools to access and study marine organisms and ecosystems. Knowledge of marine life is expanding rapidly as new species are discovered and as the complexity and biodiversity of marine organisms and ecosystems is better recognised. Marine bioresources hold great potential as a source of novel products and processes yet remain largely untapped. The application of biotechnology to these resources may help to address the global challenges of food, energy security and health and contribute to green growth and sustainable industries. At the same time, the conditions for maintaining a sustainable relationship between the conservation and use of marine bioresources is becoming better understood.

Marine organisms live in a vast interconnected system of oceans that contribute to the regulation of the planet's temperature and atmospheric conditions. Ocean currents carry marine life, nutrients and wastes within and across national borders. This fluid, shared environment and the mobile and geographically dispersed organisms contained within it present governance challenges that relate both to access to and development of marine resources. The interrelation of marine bioresources in complex marine ecosystems raises further challenges for studying marine life and for its development and conservation.

This report considers the tension between productivity and sustainability faced by those seeking to realise the potential of marine biotechnology. It seeks to identify the field's potential and the support required to achieve that potential, and explores the challenges to sustainable development, in view of the unique features of the marine environment and the need to maximise its integrity and sustainability for future generations.

The application of biotechnology to marine organisms raises several issues that create challenges for policy makers. Two recurrent themes have appeared important for the sustainable development of marine biotechnology: the need for communication among stakeholders and the need for internationally co-ordinated action. While these themes are not unique to the field, their emergence from this study's focus on the distinctive features of marine biotechnology makes them especially noteworthy and in need of attention in future policy work. As governments work to develop appropriate frameworks for marine biotechnology, it will be important to consider the main messages that emerge from this report:

- Advances in science and technology are providing new insights into the diversity of life in the oceans and contributing to a greatly improved understanding of marine bioresources. As knowledge of these resources increases, it is becoming clear that the biodiversity of the ocean offers manifold possibilities for development and exploitation.
- The biodiversity of the marine environment may help to address some global challenges through innovative food production systems, new sources of renewable energy, products for health and well-being and sustainable industries. Many countries have recognised this potential and are integrating marine biotechnology into national bioeconomy and innovation strategies.
- Strong marine biodiversity underpins healthy marine ecosystems, which contribute positively to the healthy functioning of the planet. Development of ocean bioresources must therefore occur in a manner that conserves the ocean's biodiversity. This will require a governance framework that will enable the development of marine bioresources in a sustainable manner.
- The interconnectedness of the world's oceans and their complex, widely distributed ecosystems suggests that governance of marine bioresources will be most effective if considered at international level. Several international conventions and agreements provide a framework for the conservation of marine biodiversity and for access to and sharing the benefits of marine bioresources. Yet these agreements tend to break down in areas beyond national jurisdictions; this suggests the need for finding ways to address this gap.
- Better R&D infrastructure and platforms will be needed to improve our understanding of marine bioresources and to improve our access to and development of these resources. The shared and dynamic nature of ocean bioresources, together with the sheer size of the development opportunities (and challenges), means that national and international co-operation will be needed to develop infrastructure to support marine biotechnology.
- As marine biotechnology becomes a focus of government innovation strategies and investment, it will be important to measure the return on government and private-sector investment. New measures and indicators may be needed to measure the impact of investment and the effectiveness of government policies.

- Marine biotechnology has applications in sectors such as energy (e.g. algal biofuels), pharmaceuticals (e.g. novel antibacterials), food (e.g. genomics of major food fish species) and chemical industries (a host of chemical types, such as polysaccharides). They will require different types of industry incentives and partnership strategies to foster the effective development and diffusion of technology. All sectors will benefit from discussions and engagement with relevant stakeholders at an early stage.

- Characterising and monitoring the health of marine ecosystems at many levels requires new tools and measures. Given the shared nature and complexity of marine bioresources and marine ecosystems, such tools will be most effective if developed and applied internationally. Similarly, it will be necessary to monitor and evaluate relevant policy and governance measures.

The OECD Working Party on Biotechnology is proposing to take forward work on marine biotechnology and on its potential contribution to green growth in 2013-14. This will provide the WPB an opportunity to build on this report and, with other organisations and stakeholders, to explore further the issues raised with a view to contributing to international policy and governance on marine biotechnology.

Chapter 1

Marine biotechnology for sustainability

Advances in science and technology in the last decade have increased our understanding of ocean bioresources and renewed interest in the field of marine biotechnology. As governments recognise marine biotechnology's potential to address some of today's global challenges, they are incorporating marine biotechnology in their strategies for innovation, national prosperity, and economic and social growth. As marine biotechnology becomes a prominent feature of these strategies, it will be important to ensure that the field is developed in a sustainable manner.

A better understanding of ocean bioresources

Until relatively recently, knowledge about the diversity of marine life was virtually nil. The vast volume of the oceans[1] remains largely unexplored and still relatively little is known about the range of life it supports. Fish, shellfish, marine plants and mammals have been harvested from the ocean for millennia, but accessing and understanding the totality of the ocean's bioresources is a challenge still to be met.

Darwin's historic voyage on the *HMS Beagle* in 1831 was the start of a new understanding of ocean bioresources.[2] From then and throughout much of the 19th century, marine organisms were collected and catalogued alongside more pressing ocean mapping and exploration activities. Collecting typically took place along coastal areas and in the upper layers of the ocean, and it was not until 1864 that Norwegian researchers discovered the first "deep-sea" life form, a stalked crinoids, at a depth of 3 109 metres.

Several years later the *HMS Challenger* was modified for scientific work on marine species. The Challenger Expedition (1872-76) undertook a comprehensive mapping and sampling of the oceans, coasts and deep seas. The expedition covered almost 130 000 km and discovered 715 new genera and 4 417 new species of marine organisms. This mammoth undertaking resulted in 50 published reports, and, for the next 75 years, most of what was known about world ocean biogeography stemmed from analyses of the collections made during the expedition.

Today, owing to advances in ocean exploration technology and mega-projects such as the decade-long *First Census of Marine Life* (COML, 2010; Box 1.1), ocean geography and ocean biology (and microbiology) are better understood. The census identified more than 6 000 new species to increase the number of known marine species to 250 000 (COML, 2010). Advances in genomic technologies continue to enrich our understanding of marine bioresources but reveal, at the same time, how much remains unknown. In fact, it is estimated that nothing is known about 90% of the species in the ocean (Mora et al., 2011).

Box 1.1. The census of marine life

The result of a decade of exploration involving more than 2 700 scientists from more than 80 countries and 540 expeditions, the census has contributed substantially to our knowledge of marine bioresources and underscores the untapped potential of marine biotechnology. In its summary report, the *Census of Marine Life* states: "After all its work, the *Census* still could not reliably estimate the total number of species, the kinds of life, known and unknown, in the ocean." Extrapolating from its work, the Census estimates that there may be a million kinds of marine species[3] and tens or hundreds of millions of kinds of microbes.

Source: COML, 2010.

Box 1.2. Defining biodiversity

"Biological diversity means the variability among living organisms from all sources including, inter alia, terrestrial, marine and other aquatic ecosystems and the ecological complexes of which they are part; this includes diversity within species, between species and of ecosystems."

The Convention on Biological Diversity, United Nations, 1992, www.cbd.int/convention/articles/?a=cbd-02.

"Biological diversity is the variety and variability among living organisms and the ecological complexes in which they occur. Diversity can be defined as the number of different items and their relative frequency. For biological diversity, these items are organized at many levels, ranging from complete ecosystems to the chemical structures that are the molecular basis of heredity. Thus, the term encompasses different ecosystems, species, genes, and their relative abundance."

US Congress, Office of Technology Assessment, *Technologies to Maintain Biological Diversity, 1987.*

"In the simplest of terms, biological diversity is the variety of life and its processes; and it includes the variety of living organisms, the genetic differences among them, and the communities and ecosystems in which they occur.

Keystone Center, *Final Consensus Report of the Keystone Policy Dialogue on Biological Diversity on Federal Lands, 1991.*

"Biodiversity is the totality of genes, species, and ecosystems in a region...To understand biodiversity, one has to think like a mountain and consider not only the biotic elements of plants, animals, and other living beings, but also the patterns and processes that shape volcanoes and forests."

World Resources Institute, World Conservation Union, and United Nations Environment Programme, *Global Biodiversity Strategy, 1992,* http://pdf.wri.org/globalbiodiversitystrategy_bw.pdf

"Biodiversity is not simply the number of genes, species, ecosystems, or any other group of things in a defined area... A definition of biodiversity that is altogether simple, comprehensive, and fully operational (i.e. responsive to real-life management and regulatory questions) is unlikely to be found. More useful than a definition, perhaps, would be a characterization of biodiversity that identifies the major components at several levels of organization.

"...composition, structure, and function...determine, and in fact constitute, the biodiversity of an area. Composition has to do with the identity and variety of elements in a collection, and includes species lists and measures of species diversity and genetic diversity. Structure is the physical organization or pattern of a system, from habitat complexity as measured within communities to the pattern of patches and other elements at a landscape scale. Function involves ecological and evolutionary processes, including gene flow, disturbances, and nutrient cycling."

Noss, 1990

Ocean biodiversity: Tapping an immense genetic potential

The ocean's vast size (it covers more than two-thirds of the planet's surface) and long evolutionary history are the basis of a biodiversity which dwarfs that of terrestrial environments. Life began in the ocean more than four billion years ago, and for about three billion years, life existed only in the ocean. This, along with the ocean's extremely challenging environments, has fostered the incredible biodiversity of the marine environment. For example, of the planet's 36 known phyla, all but two are found in the ocean and 13 are exclusively found there (Arrieta et al., 2010).

The term "biodiversity" was coined in the 1980s by Walter Rosen as a contraction of "biological diversity" and has since become a part of the scientific literature and popular culture. There is no single definition of biodiversity; it is usually defined in a broad context that includes ecosystem, species and genetic diversity (Box 1.2). It is this intrinsic association with genetic diversity that suggests the potential of marine biotechnology. Historically, understanding the biodiversity and inherent genetic potential of marine bioresources has been hindered by the fact that the vast majority of marine life cannot be easily cultured in the laboratory. It is suggested that just a fraction of 1% of marine bacteria can be cultured using existing methods, and it is clear that viruses, and bacterial and viral phages, present even greater challenges.

This situation is changing following a decade of investment in "omic" sciences[4] and related technologies. New genomic tools, such as high-throughput DNA sequencing, are being used to access and study samples that cannot be cultured. Scientific interest in marine bioresources has increased rapidly: in 1980 there were 108 publications related to marine biotechnology, but from 1994 to 1996 there were 700 publications in the United States alone (Leary et al., 2009).

Metagenomics,[5] which is used successfully in the terrestrial environment, is increasingly enabling the study of ecosystem segments in the marine environment at the molecular level and leading to new knowledge of marine organisms (Chen and Pachter, 2005; Culley et al., 2006; Kennedy et al., 2008; Ray et al., 2012). Metagenomics is being used to examine marine bioresources which cannot be cultured (Ferrer et al., 2009; Simon and Daniel, 2011). Early results attest to the diversity of marine bioresources as a significant source of new biological and chemical processes and products from which new bioactive compounds can be isolated, modelled or created (Box 1.3).

Box 1.3. Metagenomic sequencing

Sequencing of metagenomes collected from seawater samples of the Sargasso Sea yielded a total of 1 045 billion base pairs of non-redundant sequence, and 1.2 million previously unknown genes (Venter et al., 2004). In another expedition, the Global Ocean Sampling expedition generated 6.3 billion nucleotides of sequence enabling the identification of 6.12 million genes (Rusch et al., 2007). These genes, says Venter, are the design components of the future, and may one day be engineered to boost the carbon-fixing capacity of the ocean or to create fuel-producing bacteria.

In the ocean, most genetic diversity is microbial and resides in bacteria, archaea, protists (a group of mostly unicellular eukaryotes), and viruses. Indeed, life in the ocean is, by weight, mostly microbial (Hunter, 2011).[6] By number, 90% of individual biological life forms in the ocean are viruses, 8-9% are prokaryotes and the remainder are protists; macroscopic life forms such as fish and whales make an incredibly small numerical contribution (Suttle, 2005, 2007; Whitman et al., 1989). On average, there are 10 million viruses and 1 million prokaryotes in one millilitre of seawater (Hennes and Suttle, 1995).

As advances in genomics science and technology enable access to and analysis of these "unculturable" microbes, hitherto difficult to study at the molecular level, the number and diversity of marine microbes and genes available for biotechnological applications are increasing exponentially (Angly et al., 2006). This is prompting a paradigm shift in our understanding of the ocean. Ocean bioresources are no longer seen solely as a source of food but are being viewed as a vast reservoir of organisms and genes with virtually unlimited potential for development and exploitation.

Marine biotechnology

Marine biotechnology can be thought of as the use of marine bioresources as the target or source of biotechnological applications, and includes:

- Marine organisms, or parts thereof, used as feedstock (e.g. for production of food, fuel, materials, or bioactive compounds).
- Products extracted from marine organisms.
- Products developed in laboratories using knowledge of the natural processes or properties of marine organisms, including products created from marine DNA via genetic engineering or synthetic biology.

- Processes catalysed by marine organisms or derivatives thereof.
- Ecosystem services (e.g. biosensors and bioremediation).
- The application of biotechnology knowledge, e.g. to fish health and welfare (vaccines, feed, breeding).
- Understanding and mapping of ecosystems based on generic biotechnological tools and knowledge.

This broad understanding of marine biotechnology includes both traditional forms of marine biotechnology, such as aquaculture, and modern forms such as bioprospecting and marker-assisted selection of fish for culture. This report focuses on the forms of marine biotechnology largely made possible by advances in science and technology over the last two decades (Figure 1.1).

Figure 1.1. Marine biotechnology: Resource-infrastructure-innovations

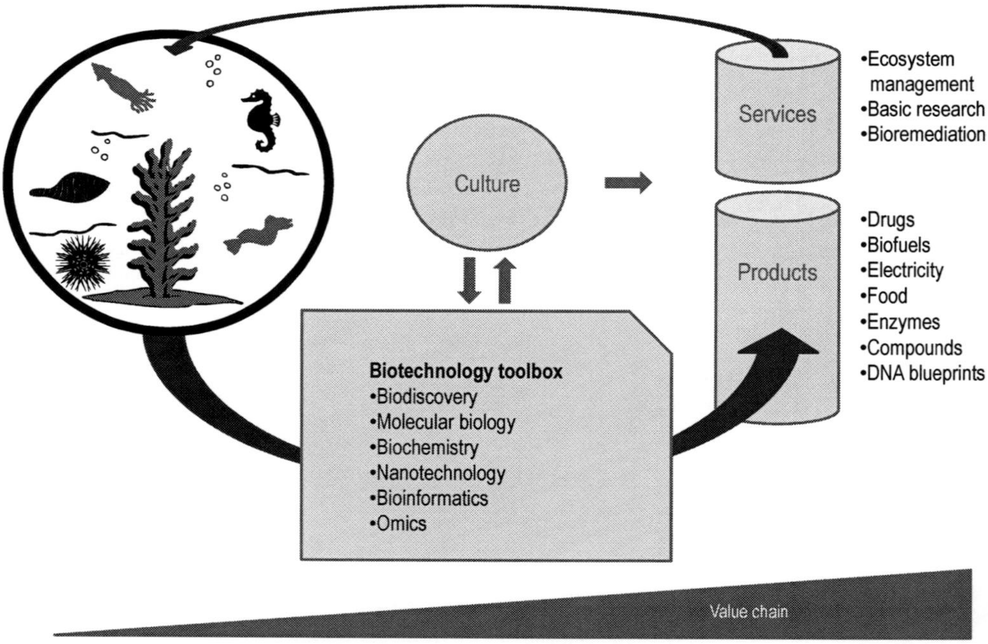

Source: Adapted from ESF (European Science Foundation-Marine Board) (2010), "Marine Biotechnology: A New Vision and Strategy for Europe", Position Paper 15, ESF, www.esf.org/marineboard, accessed March 2011.

Responding to global challenges through marine biotechnology

In the last 60 years, biotechnology has yielded some notable advances in medicine, cosmetics, nutraceuticals, food production, and industrial applications such as biorefining. The applications of marine biotechnology are also wide-ranging, and many of the opportunities discussed in recent OECD work on biotechnology – production of food and biofuels (agricultural biotechnology), development of new drugs (health biotechnology), development of new materials (industrial biotechnology) and development of bioremediation technologies (environmental biotechnology) – have parallels in the field of marine biotechnology. This suggests that marine biotechnology can help to address global challenges related to food, fuel security, population health and sustainable industrial processes.

Recent advances in science and technology have raised interest in marine biotechnology as a new source of innovation and economic growth. Governments in many countries have acknowledged the importance of marine biotechnology to the economy by dedicating funds to efforts to realise its potential or by developing formal strategies for the development of marine biotechnology.

The European Union has a number of programmes to support marine biotechnology including the ERA-NETs *AMPERA, MarinERA, MARIFISH and SEAS-ERA*[7], the Framework Programme Network of Excellence *Marine Genomics Europe*, and the Joint Programming Initiative Healthy and Productive Seas and Oceans (JPI Oceans)[8].

While Ireland has a designated marine biotechnology strategy[9], most countries incorporate marine into their biotechnology, or broader technology, strategy. Norway, for example, has a long history of maritime exploration, and marine biotechnology is an integral part of its innovation and economic development strategies and an area of continuing investment.[10] Its national strategy for biotechnology includes marine biotechnology, and, in 2009, the government published a National Strategy for Marine Bioprospecting; the strategy has a horizon of 10–15 years and focuses on creating value from biodiscoveries by working through the value chain from academy to industry. The strategy is implemented by BIOTEK2021, a programme of the Research Council of Norway. The programme also aims to stimulate the development and use of marine biotechnology, as do thematic programmes concerned with the oceans (HAV) and food production (NATUROGNÆRING). Innovation Norway has a bilateral agreement with the United Kingdom and various networking activities to foster the utilisation of marine resources through the use of biotechnological methods.

Although Canada does not have a national strategy for marine biotechnology, it has a strong interest in the application of biotechnology in the marine sector. Marine biotechnology areas such as aquaculture and bioprospecting for new compounds receive support from government programmes and initiatives, e.g. the National Research Council of Canada (NRC) and Genome Canada funding for genomics-based R&D. In the United States, marine biotechnology has been a part of policy discussions since the National Science and Technology Council's *Biotechnology for the 21st Century: New Horizons* (NSTC, 1995).

Many Asian countries also place high priority on marine biotechnology. In 1996, China added it as a separate area to its State High-Tech Development Plan (863 Program of the Ministry of Science and Technology) and has increased resources for this area under the 8th and 9th Five-Year Plans. In Japan, marine biotechnology is included in its 2002 Biotechnology Strategy and is supported by a number of ministries.

In Korea, marine biotechnology is the focus of the recent Blue-Bio 2016: A Strategic Plan for Marine Biotechnology of the Ministry of Land, Transport and Maritime Affairs. The strategy places marine biotechnology at the centre of its green growth strategy and aims to reach gross national product (GNP) per capita of USD 40 000 through advances in marine biotechnology. The strategy, accompanied by significant investment and a focus on R&D in marine organisms, marine organism production technology, development of new marine materials, and conservation of the marine environment, is designed to make Korea a world leader in the field by 2016.

Threats to marine bioresources: The sustainability challenge

As interest in marine biotechnology increases, there is growing recognition that various factors threaten the viability and diversity of marine bioresources. Of these, the largest single threat is greenhouse gas emissions (GHG). The rise in greenhouse gas emissions results primarily from combustion of fossil fuels and other carbonaceous fuels including wood, coal, oil and natural gas, and it has contributed to an increase in atmospheric carbon dioxide from 280 ppm to 396 ppm[11] since the Industrial Revolution. This has contributed in turn to global warming, an increase in the Earth's average temperature of about 0.8 °C since the beginning of the 20th century (Committee on America's Climate Choices, 2011).

Global warming and associated climate changes are causing a rise in sea levels and disrupting the delicate balance of marine ecosystems and the services they provide to the planet and its inhabitants. Temperature changes and extreme weather events are predicted to alter the habitats of marine species, expanding the habitats of some and shrinking those of others, leading to local extinction of some species and a rise in invasive species as ecosystems and food chains adjust to new conditions (Cheung et al., 2009). For example, it is estimated that warmer waters have caused a 1% decline in the phytoplankton population, the source of half of the oxygen that humans and animals breathe.

The impact of GHG emissions has been mitigated by sequestration of carbon in carbon sinks, but this leads to other problems for marine bioresources. The ocean, the largest natural carbon sink, absorbs about one-quarter of the carbon dioxide released into the atmosphere each year. Since the Industrial Revolution, it has absorbed 700 billion tonnes (or 140 Pg.-C, where 1 Pg = 5 10^{15} g) of carbon dioxide. This has lowered the ocean's pH by 0.1 unit, effectively increasing ocean acidity by 30% and reducing carbonate ion concentrations (Sabine et al., 2011). This acidification has negative consequences for oceanic calcifying organisms such as corals and crustaceans, affects fragile marine ecosystems and threatens shellfish production and associated industries (Barton et al., 2012).

Other human activities also threaten ocean bioresources. Overfishing has reduced some fish stocks to near extinction, and destructive fishery practices, such as bottom trawling, have damaged the habitat of the ocean floor. Coastal development and the resulting domestic and industrial wastes continue to perturb marine ecosystems and to threaten coastal habitats in some areas. In extreme cases, agricultural pollution has resulted in hypoxia, which weakens established ocean ecosystems and sometimes leads to permanent "dead zones".

Organic pollutants such as crude oil, hydrocarbons, petroleum oil products or halogenated compounds may originate from terrestrial sources (run-off), from spillage during transport (including pipeline failure), and from other controlled or uncontrolled releases, such as the 1991 oil spill of the first Gulf War. While much attention is paid to marine oil spills, recent work suggests that these only represent some 10% of what enters the marine environment (National Research Council, 2002). These common pollutants not only negatively affect marine life, they may also have a significant negative impact on the socioeconomic well-being of coastal communities.

Solid and particulate waste, perhaps best exemplified by the Pacific Trash Vortex[12] in the central Pacific Ocean, is yet another example of anthropogenic pollution. Plastics, chemical sludge and other debris that have been trapped by ocean currents are a continuing threat to marine bioresources (GESAMP, 2010; Andrady, 2011). Tiny particles, some with toxic chemicals or heavy metals adhering to them, may be taken up by plankton and small animals and eventually move up the food chain. Solids may clog reefs and filter feeders, negatively affecting the health and well-being of these organisms. Larger solids, plastics or ropes may ensnare larger fish. Deep sea mining[13] and transport also contribute to pollution of the marine environment, damage the functioning of marine organisms or associated marine ecosystems and affect the ecosystem services of the marine environment.

These threats to ocean bioresources are real and, in most cases, the damage is irreversible. Governments, especially those that seek economic or social benefits from marine biotechnology, have a vested interest in ensuring that ocean bioresources are developed in a manner which is both productive and sustainable. This will require finding the right balance between deriving benefits from ocean resources that are spread widely across complex marine ecosystems and maximising the integrity and sustainability of those ecosystems for future generations. Recognition of this challenge has prompted the OECD to consider how best to support the development of an appropriate framework for the sustainable growth of the marine biotechnology sector.

To that end, this report explores the potential for marine biotechnology to contribute to addressing global challenges of food and fuel security, population health and sustainable industries; the challenges for developing the field in a sustainable manner; and policy issues for achieving this sustainable development.

Chapter 2 describes the potential of marine biotechnology and its possible contributions to meeting global challenges, to societal well-being and to sustainability of the planet and its marine ecosystem. Particular attention is given to the potential of marine biotechnology resulting from advances in science and technology during the last decade, and to the sustainability imperative of marine ecosystems now and in the future.

Chapter 3 discusses the contribution of marine biotechnology to the bioeconomy, exploring existing and potential markets and other economic benefits to be realised through marine biotechnology. In the current economic situation, understanding the impact of investments and maximising returns on investment are more important than ever. This chapter therefore looks at existing measures and indicators for marine biotechnology and considers what further measures and indicators may be required.

Chapter 4 considers the knowledge-based, scientific and technological infrastructure required to reap the benefits of marine biotechnology. The Human Genome Project focused much investment and infrastructure development on human genetics. In other fields, large national or international projects have drawn financial and political attention to the infrastructures required to meet their goals. This chapter asks: "What type of infrastructure is necessary to drive development of the field? And what policies might be required to achieve this goal?"

Chapter 5 explores the nexus between science, industry and society in order to understand the conditions that will best enable successful development of marine biotechnology at regional, national and international scale. It looks at the roles and responsibilities of the private sector, government and citizens, and the interactions between stakeholders.

Chapter 6 returns to the themes discussed in Chapter 1 to present conclusions and policy considerations for enabling the sustainable development of marine biotechnology. It identifies the policy areas of greatest potential impact and highlights the challenges of supporting both productivity and ocean sustainability. It provides a foundation for future work at the OECD, and elsewhere, to explore how best to overcome the challenges and realise the potential of marine biotechnology.

Notes

1. O'Dor et al. (2009) adopt a "one ocean" perspective, recognising the interconnected nature of the world's oceans.
2. As Keynes (2000) wrote: "The first observation in his *Zoology Notes*, dated 6 January 1832, was concerned with luminous matter in the sea. [Charles Darwin's] collecting began in earnest on 10 January, when having quickly constructed the plankton net of which he drew a sketch, 'it brought up a mass of small animals, & tomorrow I look forward to a greater harvest'. The captures described in his notes were some medusae, including a Portuguese man-of-war whose powerful toxin he inadvertently got on to his fingers and into his mouth; some salpa; and 'a very simple animal' that was new to him, and remained unclassified until he returned to England".
3. A species is a group of organisms which share certain characteristics. There is much debate among biologists about how to define and identify species. Most definitions vary in the terms (or characteristics) used to group and thus define the organisms. Many biologists use Mayr's definition which describes "groups of actually or potentially interbreeding natural populations, which are reproductively isolated from other such groups" (Queiroz, 1995). Self-replicating microbes would not be considered a species under this definition.
4. Here "omic sciences" include genomics, transcriptomics, proteomics, metabolomics, epigenetics, etc.
5. Metagenomics can be defined as the application of modern genomics techniques to the study of communities of microbial organisms directly from their natural environments, bypassing the need for isolation and lab cultivation of individual species (Chen and Pachter, 2005).
6. The total mass of all prokaryotes in the ocean has been estimated at 5 200 megatonnes. Viruses are the second largest component by weight with a total mass of 200 megatonnes (Wilhelm and Suttle, 1999).
7. www.cid.csic.es/ampera/,www.esf.org/research-areas/marine-sciences/framework-programme-activities/marinera.html, www.marifish.net/, www.seas-era.eu/np4/homepage.html.
8. www.jpi-oceans.eu
9. Ireland's Sea Change Strategy (2007-13), www.marine.ie/home/SeaChange)

10. www.oslo.teknopol.no/English/MainMenu/news2/Newsletters/Oslo-Bio/Arkiv/Oslo-Bio-Update-December-2011/Marine-biotech-continues-to-make-waves/.

11. www.esrl.noaa.gov/gmd/ccgg/trends/, accessed July 2012.

12. http://marinedebris.noaa.gov/info/pdf/patch.pdf.

13. In 2012, a Canadian company, Nautilus Minerals, initiated the world's first commercial deep-sea mining project, called Solwara 1, to extract gold and copper from the floor of the Bismarck Sea in Papua New Guinea, www.nautilusminerals.com/i/pdf/SL01-NSG-DEV-RPT-7020-001_Rev_1_Golder_Resource_Report.pdf.

References

Andrady, A.L. (2011), "Microplastics in the Marine Environment", *Marine Pollution Bulletin* 62, 1596-1605.

Angly, F.E., B. Felts, M. Breitbart, P. Salamon, R.A. Edwards et al. (2006), "The Marine Viromes of Four Oceanic Regions", *PLoS Biology* 4(11): e368.

Arrieta, J.M, S. Arnaud-Haond and C.M. Duarte (2010). "What lies underneath: Conserving the oceans' genetic resources", *Proceedings of the National Academy of Sciences* 107(43):18300-18305.

Barton A, B. Hales, G.G. Waldbusser, C. Langdon and R.A. Feely (2012). "The Pacific oyster, *Crassostrea gigas,* shows negative correlation to naturally elevated carbon dioxide levels: Implications for near-term ocean acidification effects", *Limnology and Oceanography.* doi: http://dx.doi.org/10.4319/lo.2012.57.3.0698.

Chen, K and L. Pachter (2005). "Bioinformatics for Whole-Genome Shotgun Sequencing of Microbial Communities", *PLoS Computational Biology* 1(2): e24.

Cheung W.W.L., V.W.Y. Lam, J.L. Sarmiento, K. Kearney, R. Watson and D. Pauly (2009), "Projecting global marine biodiversity impacts under climate change scenarios", *Fish and Fisheries* 10:235-251.

COML (Consensus on Marine Life) (2010). *First Census of Marine Life 2010: Highlights of a Decade of Discovery*, COML, Washington DC.

Committee on America's Climate Choices, National Research Council (2011). *Causes and Consequences of Climate Change. America's Climate Choices*. The National Academies Press. Washington, DC.

Culley A.I., A.S. Lang and C.A. Suttle (2006), "Metagenomic Analysis of Coastal RNA Virus Communities", *Science* 312 (5781):1795-1798, 23 June.

ESF (European Science Foundation-Marine Board) (2010), "Marine Biotechnology: A New Vision and Strategy for Europe", Position Paper 15, ESF. www.esf.org/marineboard, accessed March 2011.

Ferrer, M., M.E. Guazzaroni, A. Beloqui and P.N. Golyshin (2009), "Metagenomics for mining new genetic resources of microbial communities", *Journal of Molecular Microbiology and Biotechnology* 16:109-123.

GESAMP (2010). IMO/FAO/UNESCO-IOC/UNIDO/WMO/IAEA/UN/UNEP Joint Group of Experts on the Scientific Aspects of Marine Environmental Protection. T. Bowmer and P.J. Kershaw (eds.). Proceedings of the GESAMP International Workshop on plastic particles as a vector in transporting persistent, bio-accumulating and toxic substances in the oceans. *GESAMP Rep. Stud.* No. 82.

Hennes, K.P. and C.A. Suttle (1995). "Direct counts of viruses in natural waters and laboratory cultures by epifluorescence microscopy", *Limnology and Oceanography*, 40:1050-55.

Hunter, P. (2011), "Massing Life", *EMBO Reports* 11(7):511–514.

Kennedy, J., J.R. Marchesi and A.D.W. Dobson (2008), "Marine metagenomics strategies for the discovery of novel enzymes with biotechnological applications from marine environments", *Microbial Cell Factories* 7:e27.

Keynes, R. (2000), "Charles Darwin's zoology notes & specimen lists from H.M.S. Beagle", Cambridge University Press, Cambridge. http://darwin-online.org.uk/content, accessed March 2011.

Leary, D., M. Vierros, G. Hamon, S. Arico and C. Monagle (2009), "Marine genetic resources: A review of scientific and commercial interest", *Marine Policy*, 33:183-194.

Mora, C., D.P. Tittensor, S. Adl, A.G.B. Simpson and B. Worm (2011), "How Many Species Are There on Earth and in the Ocean?", *PLoS Biology* 9(8): e1001127.
doi: http://dx.doi.org/10.1371/journal.pbio.1001127.

National Research Council (2002). *Oil in the Sea III: Inputs, Fates and Effects*, National Academy Press, Washington DC.

Noss, R. (1990), "Indicators for Monitoring Biodiversity: A Hierarchical Approach", *Conservation Biology* 4(4):355-364.

O'Dor R.K., K. Katja Fennel and E.V. Berghe (2009), "A one ocean model of biodiversity", *Deep Sea Research Part II: Topical Studies in Oceanography* Vol. 56, Issues 19–20:1816–1823.

Queiroz, K. de (1995), "Ernst Mayr and the modern concept of species", *Proceedings of the National Academy of Sciences* 102:6600-6607.

Ray, J., M. Dondrup, S. Modha, I.H. Steen, R.A. Sandaa and M. Clokie (2012). "Finding a Needle in the Virus Metagenome Haystack - Micro-Metagenome Analysis Captures a Snapshot of the Diversity of a Bacteriophage Armoire", *PLoS ONE* 7(4):e34238.

Rusch, D.B., J.W. Willamson, S. Yooseph, A.L. Halpern, K.B. Heidelberg et al. (2007), "The Sorcerer II Global Ocean Sampling Expedition: Northwest Atlantic through Eastern Tropical Pacific", *PLoS Biology* 5:e77.

Sabine, C.L., R.A. Feely, R. Wanninkhof, T. Takahashi, S. Khatiwala and G.H. Park (2011), "The global ocean carbon cycle", in "State of the Climate in 2010, 3. Glob. Oceans". *Bulletin of the American Meteorological Society* 92: S100–S108.

Simon, C. and R. Daniel (2011), "Metagenomic analyses: past and future trends", *Applied Environmental Microbiology* 77:1153-1161.

Suttle, C.A. (2005), "Viruses in the sea", *Nature* 437:356-351.

Suttle, C.A. (2007), "Marine viruses: Major players in the global ecosystem", *National Review of Microbiolgy,* 5(10):801-12, October.

US National Science and Technology Council (1995), *Biotechnology for the 21st Century: New Horizons.*

Venter, J.C., K. Remington, J.F. Heidelberg, A.L. Halpern, D. Rusch, J.A. Eisen et al. (2004), "Environmental Genome Shotgun Sequencing of the Sargasso Sea", *Science* 304(66): 66-74.

Wilhelm, S.W. and C.A. Suttle (1999). "Viruses and Nutrient Cycles in the Sea", *BioScience* Vol. 49, No. 10 (October):781-788.

Whitman, W.B., D.C. Coleman and W.J. Wiebe (1989), "Prokaryotes: The unseen majority", *Proceedings of the National Academy of Sciences,* 95(12):6578–6583, 9 June.

Chapter 2

The promise of marine biotechnology: Benefits for people and the planet

Recent advances in our understanding of marine bioresources have enabled better understanding of the potential contribution of marine biotechnology to social and economic growth and prosperity. Governments investing in marine biotechnology have recognised the potential for marine biotechnology to help sustain the ecosystem services the ocean provides to the planet. This chapter discusses the potential socioeconomic contribution of marine biotechnology and the importance of marine biotechnology to environmental sustainability.

Food security: Securing a sustainable food supply for a growing population

As the global population increases, demand for food and for new sources of protein is projected to grow, challenging existing food production systems. To feed the 9 billion people predicted in 2050, food output must increase by 70%.[1] This will be difficult. The reasons include climate change, urbanisation, changes in consumer tastes, scarcity of natural resources such as land and water, biodiversity issues, and the scale of the investment required to transform food production systems. Globally, consumption of animal protein is expected to double in the first half of this century. The strongest growth is expected in farmed fish and chicken, which also seem to be the sources of animal protein with the smallest carbon footprint (Nutreco, 2010).

Over 1 billion people worldwide rely on fish as their primary source of protein. Rising demand is driving innovation in fish production, as 75% of capture fish stocks[2] are depleted from overfishing. Aquaculture now produces 50% of the world's food fish (Browdy et al., 2012; FAO, 2011); it is also the fastest-growing food production sector, providing new opportunities for food production from the sea and on land and reducing pressures on wild fish stocks. So great are the productivity increases in aquaculture that it is referred to by some as a "blue revolution" that promises to transform food production as the "green revolution" in agriculture did a century earlier (FAO/NACA, 2012).

The benefits arising from the rapid growth in aquaculture have been accompanied by serious environmental, social and production challenges.[3] Reliance on fish feeds remains an issue in most countries as they are often derived from scarce wild resources. The social impact of aquaculture is multi-faceted: it creates new job opportunities but can also mean the end of traditional jobs and socially valued skills. There are also constant challenges in terms of fish health, rearing and containment. To grow and fulfil the promise of a blue revolution, aquaculture will need to balance its long-term environmental sustainability with its present goal of growing large fish rapidly.[4] Marine biotechnology may help to achieve and reconcile these two imperatives.

Marine biotechnology, in the form of new vaccines and molecular-based diagnostics, has already helped to increase production, reduce the use of antibiotics and improve fish welfare (Sommerset et al., 2005). In many places, the use of antibiotics has plummeted. In Norway 99% of farmed salmon are produced without the use of antibiotics (Figure 2.1). In other countries, however, especially developing countries without access to molecular-based tools and technologies, use of antibiotics remains widespread (Cabello, 2006).

Figure 2.1. The decline of antibiotic use in Norwegian salmon farming

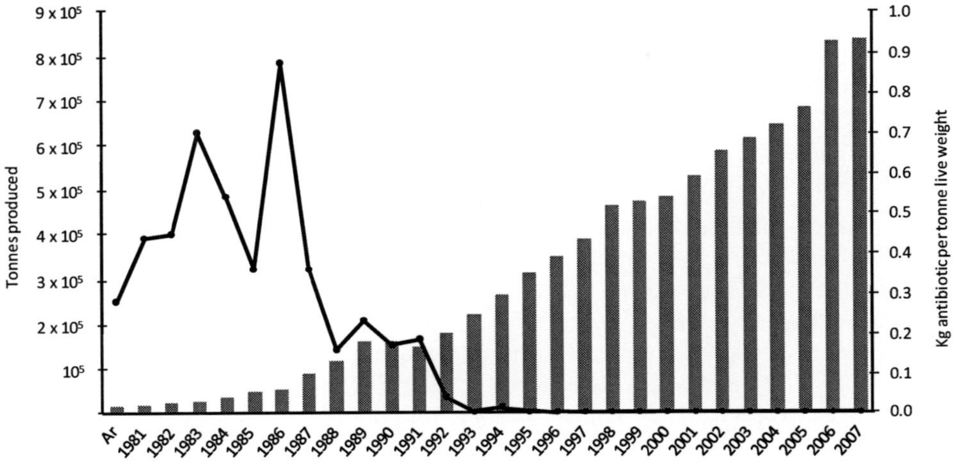

Source: Petter Arnesen (Marine Harvest ASA) at the OECD Global Forum on Marine Biotechnology: Enabling Solutions for Ocean Productivity and Sustainability, held in Vancouver, Canada, 30-31 May 2012.

The application of new genomic knowledge and technologies to the practice of aquaculture is termed "molecular aquaculture" to help to distinguish it from the more production-oriented activities in aquaculture such as improved feeding systems, cage design and husbandry.[5] Molecular aquaculture is characterised by the incorporation of new "omic" knowledge, high-throughput genomics technologies and recombinant DNA technology. These technologies have facilitated selective breeding for economically important traits such as body shape or disease resistance.

Whole-genome knowledge arising from genome sequencing projects (such as for cod or Atlantic salmon) is providing new inputs for marine biotechnology and new opportunities for aquaculture and wild stock management regimes. Genomic knowledge is being used to study species which are not currently the focus of large-scale cultivation efforts in order to identify new species for culture. Genomics is improving understanding of these species – their life cycle, nutritional requirements, pathogen susceptibilities – and providing a basis for developing improved feeds (e.g. less reliant on fish oils), production methods and fish health tools.

Genomic and related technologies have also been used to create new DNA-based vaccines for economically important diseases (e.g. Apex®-IHN, Novartis, for the treatment of infectious hematopoietic necrosis in farmed salmon) and highly sensitive specific tools for disease detection (Cunningham, 2002).

More controversially, recombinant DNA technology has been used to modify fish genetically. Like genetically modified (GM) crops, fish may be modified by the inclusion of genes from other species to improve productivity traits. This may prove to be a way forward in at least some jurisdictions. In the United States, the Federal Drug Administration (FDA) is moving closer to approving the first GM salmon, which contains a growth hormone gene from a related species which allows the salmon to grow to market size in half the normal time.

Molecular aquaculture holds great potential for increasing sustainable food production to meet anticipated increases in global demand through the culture of species such as salmon, tilapia, shrimp and oysters. However, molecular aquaculture is developing and diffusing at different rates in different countries, potentially limiting the productivity gains and sustainability of the endeavour. Developing countries face challenges for accessing new technologies and financial capital while more developed countries face challenges associated with public support and public-private partnerships. These challenges will need to be addressed if molecular aquaculture is to reach its potential.

Health: Biomedical, pharmaceutical and nutraceutical applications of marine biotechnology

Countries' grand challenges in terms of health and well-being vary (Varmus et al., 2003; Daar et al., 2007; World Health Organization, 2012; http://grandchallengesgmh.nimh.nih.gov/). In developing countries, infectious disease and the cost of treatment and vaccination regimes are leading challenges, while developed countries must deliver appropriate services to ageing populations and face increasingly stratified disease and greater incidence of drug resistance. As the world's population is ageing and growing, these challenges will only increase, but pharmaceutical companies struggle to meet new demands. To improve the health of the world's citizens, there is a need for new and more effective drugs and natural products to support citizens' health and well-being.

Since the identification of bioactive nucleosides from marine sponges over five decades ago (Bergman and Freeney, 1950, 1951; Bergmann and Burke, 1955), well over 20 000 novel marine natural products[6] from marine organisms have been discovered (Hu et al., 2011). Some marine organisms contain, or produce, bioactive or structural compounds that can be used to manage pain or reduce inflammation, to treat cancer or other diseases, as new materials for dressing wounds, or to regenerate tissue. Marine sponges or symbiotic microbes have been used as sources of products, as have fungi and, increasingly, marine bacteria.

Chitin, obtained from the shells of crabs, lobsters and the internal structures of other invertebrates, has anti-bacterial, anti-fungal and anti-viral properties which make it attractive for use in medical materials such as wound dressings and surgical sutures (Jayakumar et al., 2010). Certain siliceous sponges are of interest because of their ability to form silica skeletons. Silica from these organisms and, increasingly, the blueprints for silica skeleton formation, are finding use in a range of biomedical applications such as coatings for metal implants used in surgery, adhesives for drug delivery, and microelectronic fabrication (Andre et al., 2012). Similarly, the agar derived from macroalgae is being used in drug encapsulation, and collagen-based marine sponge skeletons have potential uses in bone repair (Zheng et al., 2007).

The isolation of bioactive arabinonucleosides from the sponge *Tethya crypta* in the 1950s led to the development of two synthetic drugs, Ara-C (against leukaemia) and Ara-A (for treating viral infections). Despite this promising start, it was not until 2004 that the next marine-derived drugs, ziconotide (Prialt®), isolated from cone snails, and trabectedin (Yondelis®), isolated from sea squirt, were commercialised. The conotoxin in cone snail venom is usually lethal to humans, but in small quantities it can be a useful anaesthetic or analgesic or be used in drugs for conditions such as epilepsy and psychiatric disorders. Today, there is strong commercial interest in conotoxins, and 251 patents and patent applications have the term "conotoxin" in the title.[7]

Figure 2.2. Trends in novel products obtained from marine organisms (number of products)

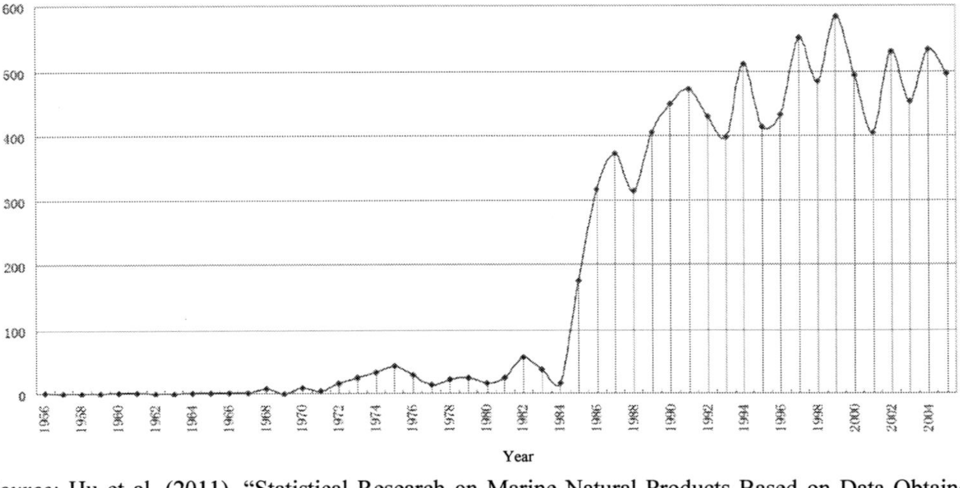

Source: Hu et al. (2011), "Statistical Research on Marine Natural Products Based on Data Obtained between 1985 and 2008", *Marine Drugs* 9(4): 514–525. doi: http://dx.doi.org/10.3390/md9040514.

Genomics are providing new insights into the genetic diversity of marine bioresources and revealing new sources of drugs (Trincone, 2011). As a result, the number of promising marine-derived compounds or secondary metabolites is increasing rapidly and some are already in the drug development pipeline (Figure 2.2). From 1998 to 2006, the pipeline included 592 marine compounds with anti-tumour and cytotoxic activity, and 666 additional chemicals with pharmacological activity (anti-bacterial, anti-coagulant, anti-inflammatory and anti-fungal, as well as effects on the cardiovascular, endocrine, immune and nervous systems) (Mayer et al., 2010).[8]

The marine-related pre-clinical pipeline for drug development is growing and becoming truly global; investigators from 32 countries were involved in 2007-08 alone. By 2010, there were over 36 marine-derived drugs in clinical development, including 15 for cancers. Close to half of all current anticancer discovery efforts focus on marine organisms. Two years later, seven marine derived drugs had received FDA approval, eleven drugs were in clinical testing and 1 458 were in the clinical pipeline.[9]

Marine microbes, and bacteria in particular, are the focus of much attention, as they are increasingly seen as a particularly rich source of bioactive compounds (Gokulkrishnan et al., 2011). This is due both to the complex nature of marine ecosystems and to estimates of undiscovered bacteria in the marine environment.[10] As most existing drugs are derived from terrestrial sources, marine resources, particularly marine microbes, are a largely untapped resource (Chin et al., 2006; Newman and Cragg, 2007).

Marine biotechnology may thus make significant contributions to the development of new antibiotics, anticancer and immune system modulators. Antimicrobial resistance due to widespread use of antibiotics for human health and agriculture is a serious health threat and is the focus of increasing public and government concern. The World Health Organization (WHO) has identified this as one of the three main threats to human health. The problem is likely to get worse as there are few new candidate drugs in development; bacteria are becoming resistant to antibiotics faster than effective replacements can be developed (Dwyer et al., 2009).

Pharmaceutical development relies more on exploitation of new compounds discovered through metagenomics and screening of marine biobank samples for bioactivity. The major bottlenecks in the marine pharmaceutical pipeline include insufficient funding for basic marine pharmacology and technical challenges for the characterisation of unknown taxa and gene functions. Several groups have R&D programmes to develop novel antibiotics through the isolation and characterisation of potent substances from the sea. In 2006, the Scripps Research Institute launched a programme of antimicrobial R&D running from initial discovery to development and testing to clinical trials.

Table 2.1. Functional food from marine biomass

Ingredient	Studies
	Fish
LC n-3 PUFA (omega-3 fatty acids)	The metabolic syndrome, cancer, Inflammatory diseases, brain function(dementia and macular degeneration / schizophrenia / depression), effects during pregnancy
Marine phospholipids	Therapeutic effect on brain, suppression of cancer
Vitamin D	Bone health, Inflammatory diseases, cancer, brain, pregnancy
Selenium (Se)	Immune system, viral infections, reproduction, thyroid function, mood, cancer, mammary gland (rats),colon cancer
Fish peptides and hydrolysates	High blood pressure, low immune response, cancer anaemia
Selected amino acids in fish	Atherosclerosis, blood lipids, inflammation, oxidative stress, diabetes II
Fish proteins	High blood pressure, lipid metabolism, obesity/metabolic syndrome, glucose and lipid metabolism, insulin sensitivity
	Shellfish
Chitosan and glucosamine	High cholesterol, infection, cancer, low immune response, wounds, Alzheimer's disease
Chondroitin sulphate	Osteoarthritis, obesity/weight loss, cancer, oxidative stress, neuro-related diseases
LC n-3 PUFA fortification	Preterm infants, term infants, fortification increases intake, lipid peroxidation
	Seaweed
Proteins, peptides and amino acids	High blood pressure, low immune response obesity/metabolic syndrome, glucose and lipid metabolism
Fatty acids	Heart diseases, inflammatory diseases
Polysaccharides Sulphated fucan Sulphated galactan	Oxidative stress, virus, cardioprotective
Metabolites Polyphenols Steroids	LDL cholesterol, valuable curative properties, anti-oxidant, anti-diabetic, anti-inflammatory
Vitamins Vitamin C Vitamin E	Oxidative stress
Pigments Carotenoids Chlorophylls	Cerebro-vascular diseases, metabolism, obesity, diabetes

Marine bacteria are, however, not the only potential source of new drugs. Other marine microbes, aquatic plants and larger marine organisms are also a focus of pharmaceutical research. The discovery and development of novel bioactives from marine sources is accelerating owing to recent advances in high throughput screening and metagenomic analysis, and compounds from harmful algal blooms are showing promise in pharmaceutical terms (Waters et al., 2010).

Nutrients, enzymes, metabolites and other compounds from marine bioresources are also contributing to nutraceutical applications and the development of functional foods. Macroalgae, fish and even bacteria are used as sources of essential fatty acids, including arachidonic acid (ARA) and docosahexaenoic acid (DHA). Vertebrates and shellfish are good sources of calcium or chitin (and derivatives such as glucosamine) which have found application as nutritional supplements or aids. Marine organisms also produce a number of metabolites and active compounds that can be incorporated in a range of nutraceuticals containing active ingredients such as antioxidants, essential oils and vitamins that support good health (Table 2.1).

As functional foods are largely biomass-based, they require relatively less investment and research intensity than pharmaceuticals. Documentation of the effects of functional foods/nutraceuticals is a neglected area but has the potential to add much value to these products. Functional foods have relatively strong market penetration and public acceptance, but their commercialisation will require the development of sustainable culture, capture or harvesting as well as appropriate extraction and preservation methods.

Fuel security: The use of marine organisms to produce sustainable and renewable energy

Most OECD and many non-OECD economies are committed to reducing their carbon footprint. Many OECD members are turning to renewable biomass to supplement, and perhaps eventually replace, some petroleum-based feedstock. For different countries, this move may be driven by issues relating to fuel security or economic vulnerability.

Biomass is a biological source of organic carbon (e.g. wood, plants, animal matter and a variety of organic solids including industrial and agricultural waste) which can be used directly or converted into energy products such as biofuel and biogas. Biomass is therefore a renewable source of energy, and conversion of biomass into biofuel is one instance of agricultural or industrial biotechnology.

First-generation biofuels, such as biodiesel made primarily from grains, seeds or commodities such as maize and sugar cane, are produced in many parts of the world. These biofuels have been criticised for diverting food sources away from the human food chain and for creating unsustainable patterns of land use.[11] Second-generation biofuels, also called advanced biofuels, are the same end product but are considered more sustainable. They are derived from sustainable sources (e.g. non-food crops and waste biomass) and/or have a lower carbon footprint than first-generation biofuel feedstocks. They therefore lessen the concerns raised by the latter. Although they raise fewer sustainability issues, they create development and commercialisation hurdles, largely owing to the difficulty of extracting the useful sugars locked in the fibrous biomass.

In 2008, the OECD published an economic assessment of second-generation biofuel support policies (OECD, 2008). It concluded that government support of biofuel production in OECD countries is costly, has a limited impact on reducing greenhouse gases and improving energy security, and has a significant impact on world crop prices. The report concluded that other forms of bioenergy, such as bioheat, biopower and biogas, would represent more economically viable and environmentally sustainable ways to reduce greenhouse gases.

To become economically and environmentally viable, biofuel production will need to address technical and commercial challenges such as treatment or disposal of by-products, carbon neutrality, production and capital costs, scale-up, and integration with existing infrastructure (Coyle, 2010). Algal biofuels, also known as third-generation or next-generation biofuels, may address some of these challenges and make biofuels more economically viable and environmentally friendly.

Algae biomass may be composed of either microalgae or macroalgae (seaweed). Algal biofuels result from the application of marine biotechnology to algae biomass to generate biodiesel, bioethanol, biogasoline, biomethanol, biobutanol and other biofuels. Algae biomass offers many of the advantages associated with first- and second-generation cellulosic biomass and lacks lignin, a plant material whose presence in cellulosic biomass presents significant processing challenges. This is the most technologically significant advantage of algal biomass for biofuels. Other advantages of algae biomass include the fact that its production is not geographically limited and its cultivation can avoid competition with food production. Adoption of algal biofuels can rely in part on investments and infrastructure developed to accommodate first- and second-generation biofuels.

The use of algae for biofuels is attractive for many reasons. Algae may produce more energy/tonne and may be grown more quickly (typically in 1-10 days) than conventional crops such as soybean or cotton (Potters et al., 2010). Theoretical calculations of biofuel production vary, yet production of ethanol from algae is widely considered to exceed production from terrestrial crops (Scott et al., 2010; Tan et al., 2011) (Table 2.2). Cultivation of microalgae may also have a smaller physical footprint than comparable land-based biomass. It has been estimated that no more than 39 000 km^2 of algae, an area corresponding to the Sea of Azov or to around 10% of the land surface of Germany, would be sufficient to replace all the petroleum fuel in the United States (Potters et al., 2010).

While algal biofuels require fertilisers or organic material, these can be carbon- or nitrogen-rich waste gas streams and provide an opportunity to biofix these greenhouse gases before they enter the environment (Mussatto et al., 2010; Ho et al., 2011). Algae can be produced using ocean and waste water, obviating the need for fresh water. Microalgae can also be produced on land, and algae are biodegradable if released into the environment as waste (Christenson and Sims, 2011).

Table 2.2. Theoretical calculations of biofuel production from different crops

Crop	Oil yield ($l\ ha^{-1}$)	Land area needed [1] (M ha)	% of existing US cropping area
Corn	172	1 540	846
Soybean	446	594	326
Canola	1 190	223	122
Jatropha	1 892	140	77
Coconut	2 689	99	54
Oil palm	5 950	45	24
Microalgae[2]	136 900	2	1.1
Microalgae[3]	58 700	4.5	2.5

1. To meet 50% of all transport fuels needs in the United States. 2. 70% oil (by weight) in biomass. 3. 30% oil (by weight) in biomass.

Source: T. Tan, J. Yu and F. Shang (2011), "2.58 - Biorefinery Engineering", in *Comprehensive Biotechnology* (Second Edition), Vol. 2, pp. 815-828.

Microalgae may be grown in ponds or photobioreactors and may be used as feedstock for several renewable fuels. Marine biotechnology clearly has a role to play in the successful development of algal strains. The biochemical composition of algae may be modified by altering bioreactor-based growing conditions to produce valuable co-products such as proteins and residual biomass. Growth conditions and extraction processes need to be further developed before algal biofuel production will be commercially viable (Day et al., 2012). The economic feasibility of these processes needs to be addressed and a full life-cycle analysis is needed to study carbon dioxide production, the biofixation ability of microalgae strains, and the stability of these strains under production conditions (Brennan and Owende, 2010; Ho et al., 2011). Further applications of marine biotechnology, potentially including the genomic modification of algal strains to suit production scenarios or the harvesting of other bioactives in the fuel extraction process, will undoubtedly be required for commercialisation on a wide scale. Adaptation of the biorefinery concept to marine raw materials offers one approach to addressing these challenges.

Macroalgae-based biofuels present similar but slightly different opportunities and challenges. Macroalgae are a readily accessible source of biomass, their culture and production methods are well developed, and the supply chain is well established. However, the physical footprint of macroalgae is larger than that of microalgae and, because it is more amenable to growth in the ocean, there are challenges in terms of containment and interaction with facilities such as fisheries and wind farms. Challenges in terms of biofuel extraction and waste are being addressed. Bio Architecture Laboratories (BAL) has recently engineered a microbe to degrade and a pathway to metabolise alginate, the most abundant sugar in seaweed (Wargacki et al., 2012). Alginic acid/alginates constitute 20-30% of the total dry matter content of brown seaweeds. The BAL platform can convert seaweed carbohydrates into a renewable chemical intermediate that is scalable and can be used to produce both fuels and a variety of chemicals for green plastics, surfactants, agrochemicals, synthetic fibres and nutraceuticals.

Although currently more expensive than other biofuels, algal biofuels represent a novel marine biotechnology which has passed the proof-of-concept stage (Mussatto et al., 2010) and is now the focus of significant activity in the private and public sectors. Apart from the technical challenges noted above, commercialisation of biofuels and biochemicals faces two major challenges. On the investment side, the slow pace of decision making and the level of crossover among funding domains impede progress; on the market externality side, biofuel companies compete against 100 years of technology development and substantial government support in the oil and gas sectors.

Critical questions remain concerning the viability of using algae for large-scale biofuel production. A recent German report (Leopoldina, 2012), for instance, argued that the oceans are unsuitable as a source of biomass for large-scale biofuel production owing to the rapid turnover of unicellular phytoplankton as a result of grazing by zooplankton. To answer questions regarding viability will require a detailed life-cycle analysis (LCA) that provides internationally acceptable performance data. While algae have many advantages, a detailed analysis may reveal further problems. For example, the use of fertiliser releases nitrogen-based GHGs with a much higher global warming potential than CO_2, and the extent of the problem will depend on the rate of application of fertiliser.

Aside from the production of renewable biofuels, marine biotechnology may be used to make the extraction of fossil fuels more efficient. Extraction of fossil fuels is often relatively inefficient because of the porosity of the rock or the viscosity of the crude oil. Only a fraction of the oil in the oil fields tapped to date has been removed, and reservoirs with a large percentage of oil are often abandoned owing to the increasing difficulties of extraction as the reservoir is exploited. Marine microbes can be used to increase the efficiency of oil recovery by decreasing the viscosity of the oil or increasing the permeability of the rock material in the reservoirs. While microbial enhanced oil recovery (MEOR) is still controversial,[12] it is another way in which marine biotechnology can contribute to addressing the global challenge of energy security (Brown, 2010).

Microalgae, macroalgae and bacteria have also shown their utility in microbial fuel cells, i.e. systems that harvest the electricity generated by microbial metabolism (Reimers et al., 2001; Girguis et al., 2010).

Industrial processing: Applications in research, manufacturing, processing and other sectors

Pressures to reduce GHG emissions and improve environmental sustainability have prompted significant investment in the development of sustainable industries as a source of green growth. Tools based on marine biotechnology can be widely used in industrial processing and manufacturing and can play an important role in global green growth efforts. Many marine organisms, or their products, including several of those mentioned above, are good sources of novel enzymes, biopolymers and biomaterials. Bioactive molecules, compounds or enzymes can be cultured or harvested directly as feedstock from marine sources or used to synthesise analogues.

For example, in addition to their fuel potential, algae and other marine biomass represent largely untapped alternatives to so-called platform chemicals and even functional food products. Microalgae are rich in polyunsaturated fatty acids (PUFA), a vegetable alternative to fish oils and oils rich in omega-3 fatty acids. They may therefore find application in fish feed and products for human consumption prior to the esterification needed for biofuel production.

Chitin, a polysaccharide consisting of units of N-Acetyl glucosamine, is a marine resource that has found wide application (Jayakumar et al., 2010). Obtained from the shells of crabs, lobsters, shrimp and the internal structures of other invertebrates, chitin and its d-acetylated form chitosan are used as stabilising agents in foods and cosmetics, to treat water for food preservation, and in antifouling applications.

Most of the 3 500 enzymes identified from microbial sources so far are from terrestrial sources. Those identified in the marine environment have tended to come from organisms living in environmental extremes and have a wide range of applications in industrial and agricultural processing.

Biosurfactants and bioemulsifiers are two important bioactive compounds derived from bacteria. These amphiphilic compounds display a range of surface activities that allow for the solubilisation of hydrophobic substrates. They are produced by a huge variety of bacteria and have a wide range of structural and functional forms. Depending on their form they have a range of potential industrial and environmental applications (e.g. emulsification, detergency, dispersion and solubilisation of hydrophobic compounds) which are of interest for replacing synthetic surfactants (Satpute et al., 2010).

Similarly, bacteria-derived exopolysaccharides,[13] high molecular weight polymers secreted by bacteria, are being considered as substitutes for synthetic or plant and algal gums (including marine-derived carrageenan). Bacteria-derived gums have properties similar to plant and algal gums and can be used for stabilisation, gelling, adhesion, thickening and many other industrial applications. For example, xanthan gum, from the bacterium *Xanthomonas campestris*, is widely used in foodstuffs and cosmetics. It is used in large quantities during drilling for oil, particularly in horizontal drilling, as a mud thickening agent with excellent rheological properties. Marine bacteria are expected to be a valuable source of new exopolysaccharides with diverse and useful functional and structural properties.[14]

Bacteria-derived silicas, in the form of novel biosilicas with unique electrical, optical and catalytic properties, also have great potential in nanomaterials. Marine-derived silica may be used as a carrier or stabiliser of manufactured products, can confer unique properties to adhesives or paints, and may be used as an insulator or filler of materials.

Marine biotechnology applications are also a source of new products for conducting biological research. Agar and agarase from macroalgae are stable for culturing, while polymerase enzymes such as *Thermus aquaticus* isolated from thermal vents in the marine environment are of use in the polymerase chain reaction (PCR) used to amplify small amounts of DNA. The bioluminescence compound aequorin and the fluorescent molecule GFP (green fluorescent protein) isolated from jellyfish have also found applications as probes and in imaging applications in life science research. Bioluminescence has a range of environmental applications as luminescent bacteria can be used to detect contaminants in wastewater and soil.

Other environmental applications of marine biotechnology cover environmental monitoring and bioremediation (discussed below) and biofouling. Biofouling can reduce the fuel efficiency of transport vessels; in aquaculture, it can reduce the functionality of nets and corrode and destabilise marine structures. Biofouling generates an economic expense, but attempts to overcome it may harm the environment. Paints designed to reduce biofouling and agents to clean biofouled materials (e.g. aquaculture nets) are often toxic and harmful to the environment. Marine biotechnology may lead to the development of naturally occurring marine organisms or derivatives that prevent biofouling. Rocks or crustaceans devoid of barnacles and other biofouling, often found in coastal regions, are an obvious source of such organisms. In addition to being eco-friendly, the products may be more effective than existing chemical products.

Marine biorefineries

There is considerable speculation regarding the business model and economics of biorefineries based on marine raw materials. The most attractive approach involves integration of all components in a global business model. In the integrated marine biorefinery, biofuel is one product but more valuable feed, food and other materials are also produced (Figure 2.3), and indeed may have to be produced to make the refinery financially viable. As in petrochemical refineries, profit margins on bulk production fuels are likely to be low, and higher revenues from high-value, low-volume materials make biorefining more attractive.

There are also opportunities for applying biorefinery-type processes to extract commercial products from microalgal biomass. Microalgal biomass cultivated for its lipid content for conversion to biodiesel offers several possibilities for obtaining additional commercial materials such as fermentation to obtain ethanol (low conversion rates) and biogas. It is also possible to produce protein-rich feed for animal and human consumption. Figure 2.4 shows some permutations of the integrated marine biorefinery concept. The design and engineering principles for marine biorefining are in their infancy compared to biorefineries for terrestrial crops.

Figure 2.3. Value pyramid for marine biotechnology products

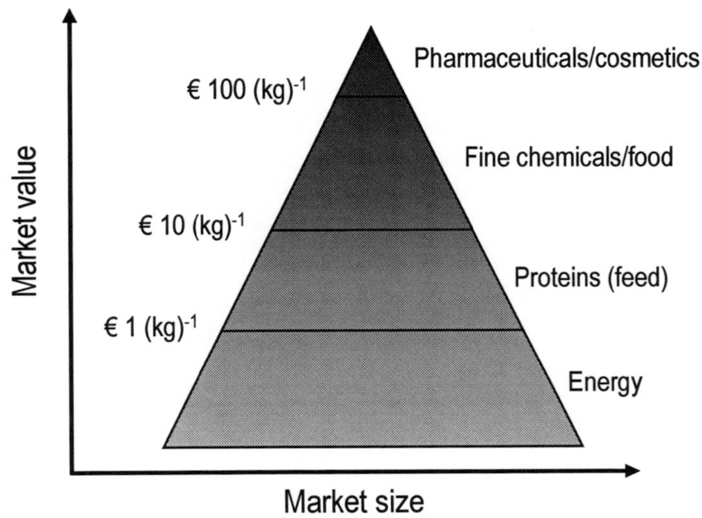

Source: Sustainable Energy Ireland (2009), "A review of the potential of marine algae as a source of biofuel in Ireland", *Sustainable Energy Ireland*, Dublin, February.

Figure 2.4. Permutations of the integrated marine biorefining concept

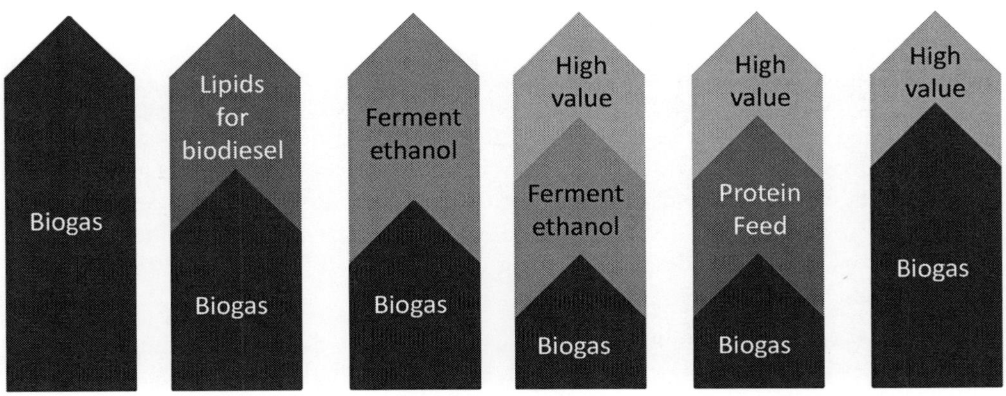

Source: Sustainable Energy Authority of Ireland (SEAI) (2009), "A review of the potential of marine algae as a source of biofuel in Ireland", *Sustainable Energy Authority of Ireland (SEAI)*, Dublin, February.

Ecosystem goods and services from marine resources

Marine bioresources provide a number of important ecosystem goods and services for the planet and its inhabitants. Costanza et al. (1997) classified 17 different ecosystem services and estimated that marine systems contribute about 63% of value (USD 20.9 trillion a year), mostly from coastal systems. Marine organisms (microalgae, fish and invertebrates) are a source of food for billions of people and livestock. The oceans are well known as regulators of global temperatures and filters of pollution. They are also sinks for carbon and nitrogen, and a source of oxygen and food.

Living in the ocean's surface water, in easy reach of light, phytoplankton are estimated to produce half of the oxygen breathed by humans and animals. In addition, they have an important place in the food web and play a valuable role in carbon cycling by locking away carbon dioxide and nitrogen, which is eventually deposited on the ocean bottom, thereby slowing the impact of global warming. The ecosystem services provided by the marine environment are attributable to its vast size and to a large extent to its complex ecosystems and biodiversity.

However, the oceans' bioresources and biodiversity increasingly face direct and indirect threats (see Chapter 1) that could disrupt marine ecosystem services. It is therefore most important to preserve the marine environment and its bioresources. Marine biotechnology can play an important role in reaching these goals.

Marine-derived biosensors can help to monitor the marine environment. Biosensors are devices composed of a biological sensing component linked to a signalling component which can reveal the presence of an element, molecule or organism of interest. They can detect changes in analytical or biological parameters and may be useful for detecting quickly the presence of invasive species that can disrupt marine ecosystems and the habitats of marine or terrestrial organisms. In 2011, a team at the Virginia Institute of Marine Science reported the creation of a portable biosensor that could detect marine pollutants, including oil, much more quickly and cheaply than current technologies (Spier et al., 2011). If deployed near oil facilities, such sensors could provide early warning of spills and leaks and track dispersal patterns in real time.

Marine biotechnology can also contribute to the preservation of the marine environment and ecosystem services through the development of DNA-based monitoring tools (Bott et al., 2010). These tools have a range of applications; they enable the detection of genetically modified organisms and aquaculture escapees, validate the identity of species, and alert environmentalists to the presence of invasive species.

Beyond monitoring of the marine environment, marine biotechnology has a role in remediation. Researchers are screening naturally occurring microbial populations in the marine environment to identify bacteria that metabolise certain types of hydrocarbons. These bacteria can be used to break down pollutants without any danger to the ecosystem as they occur naturally in the environment.

Although bioremediation has seen some promising advances, there is room for improvement. It is often slow and may not perform consistently under different environmental conditions, such as variation in temperature and wave strength. Bioremediation is often boosted by additives that enhance the activity of endogenous microbes that degrade hydrocarbon compounds (biostimulation) or by the addition of new oil-degrading bacteria (bioaugmentation). Biostimulation was used successfully during the Exxon Valdez oil spill, through the application of fertiliser to more than 120 km of oiled shoreline (Bragg et al., 1994). However, such processes may not be suitable under all circumstances and their effectiveness may vary depending on the form of hydrocarbon, volume of spill, and wind and sea conditions. Indeed, they may cause further damage to fragile marine environments. Genetic manipulation of these bacteria and/or identification of new bacteria with appropriate remediating abilities are now a focus of marine biotechnology research.[15]

Conclusions: Marine biotechnology for people and planet

The outlook for marine biotechnology has changed profoundly over the last decade in large part owing to advances in science and technology, in particular the "omic" sciences. These advances provide new insights into marine bioresources and improve the ability to access, manipulate and develop these resources to address some of today's grand challenges. Ocean bioresources are no longer viewed solely as a source of food but more as a vast reservoir of a diversity of organisms and genes with virtually unlimited potential for development and exploitation. Under the right conditions, marine biotechnology may provide new foods and food production methods, new health products, sustainable renewable energy alternatives and new sustainable industries.

The application of marine biotechnology to marine biomass – fish, shellfish, macroalgae and microalgae, related waste – provides new opportunities for improved and expanded food production, development of sustainable biofuels and new natural products for health and well-being. Perhaps the biggest challenge to biomass-based development is the need to harvest biomass in a sustainable manner. A number of contemporary examples of overfishing (e.g. cod fisheries in Newfoundland, anchovy fisheries off the coast of Peru) illustrate the economic, social and ecosystem damage that can be done by unsustainable harvesting of marine biomass. Marine biotechnology can help to ensure the sustainable use of biomass through the development of new culture, production and processing techniques and practices.

The genetic diversity of marine bioresources, in particular of marine microbes, provides a wide array of opportunities and challenges for marine biotechnology. It holds significant potential for discovering new bioactive compounds with wide applicability in drugs and in greener, more sustainable industries. The need to develop marine resources in a sustainable way remains, but, for marine microbes, it is prefaced by the need to access and characterise these genetic resources in a complex and dynamic ecosystem. Here the need is to sift through and translate huge amounts of data in order to understand and develop these resources effectively. Genomics and related tools are providing a means to access and characterise these resources, but this work will clearly become more efficient and effective with greater understanding of the complexity of marine bioresources.

Marine resources are crucial for the survival of the planet and the quality of these resources can be affected positively or negatively by the actions of many countries. Because marine bioresources are inextricably linked to the ecosystem services the ocean provides to the planet and its inhabitants, it is essential to preserve their integrity and biodiversity. Marine biotechnology can provide a means of monitoring and even remediating the marine environment through biosensor and bioremediation applications. Such applications are essential for understanding and predicting changes in this environment and for conserving and managing the coastal and marine resources that are critically important to a bioeconomy.

While the potential for marine biotechnology is clear, delivering on its promise may require the attention of governments, policy makers and others. Put simply, the challenge is to extract value from resources that are spread widely across a complex marine ecosystem while maximising the integrity and sustainability of that ecosystem for future generations. Policy work to ensure the protection and appropriate development of shared marine resources is a global imperative and actions are most likely to be effective if they are harmonised across countries.

Notes

1. www.fao.org/wsfs/forum2050/wsfs-forum/en/, and www.fao.org/fileadmin/templates/wsfs/docs/expert_paper/How_to_Feed_the_World_in_2050.pdf.
2. www.fao.org/fishery/topic/3380/en.
3. Many OECD countries have recognised the importance of sustainable aquaculture (e.g. OECD Workshop on Advancing the Aquaculture Agenda: Policies to Ensure a Sustainable Aquaculture Sector, April 15-16, 2010).
4. The development of environmental performance standards for aquaculture sites and the development of integrated multi-trophic aquaculture are helping to improve the situation (Chopin, 2010).
5. Introduced by the Co-ordinated Working Group on Marine Biotechnology under the KBBE-net, 2009, http://ec.europa.eu/research/bioeconomy/pdf/cwg-mb_to_kbbenet_report_final.pdf.
6. Chemical compound or substance produced by a living organism in nature.
7. www.espacenet.com, accessed August 2012.
8. See *The Global Marine Pharmaceuticals Pipeline* at http://marinepharmacology.midwestern.edu/.
9. www.oecd.org/sti/biotechnologypolicies/Session%204%20Mayer.pdf.
10. Products such as the polyketide bryostatin-1 first isolated from *Bryozoa* are now known to be a secondary metabolite from associated bacteria (Waters et al., 2010).
11. Brazilian biofuels derived from sugar cane waste escape the negative press coverage accorded first-generation biofuels.
12. MEOR is a controversial practice: there have been many claims of success but also failures. There is no defined standardised technology, and field trials, often using dubious products, have varied widely in quality. Trials may not have been conducted on the most appropriate wells and reservoirs, and monitoring the operations can be very time-consuming and costly. Further, the trials are often conducted during routine oilfield operations so that it is difficult to gather robust data. Nevertheless, interest in MEOR persists as it may offer an inexpensive option for EOR, and improved recovery of even a few percent can amount to large

quantities of oil and profit over the extended working life of an oil reservoir.

13. Marine plants and macroalgae are also a good source of exopolysaccharides which are used in a number of industrial applications.

14. For example, some *Bacillus* species produce exopolysaccharides with antimicrobial properties.

15. The OECD worked on bioremediation technologies as early as the 1990s, and the 2010 Rimini workshop on Environmental Biotechnology also addressed hurdles to successful uses of such applications.

References

Andre, R., N. Tahir, F. Natalio and W. Tremel (2012). "Translating the genetically controlled mechanism of biosilification to the bioinspired synthesis of multifunctional inorganic and hybrid materials", *FEBS Journal* 279:737-749.

Bergmann, W. and R.J. Feeney (1950). "The Isolation of a New Thymidine Pentoside from Sponges", *Journal of the American Chemical Society* 70:809-2810.

Bergmann, W. and R.J. Feeney (1951), "Contributions to the study of marine products. XXXII. The nucleosides of sponges", *Journal of Organic Chemistry* 16:981-987.

Bergmann, W. and D.C. Burke (1955). "Contributions to the study of marine products. XXXIX. The nucleosides of sponges. III. Spongothymadine and spongouridine", *Journal of Organic Chemistry* 20:1501-1507.

Bott, N.J., K.M. Ophel-Keller, M.T. Sierp, K.P. Rowling, A.C. Mckay, M.G.K. Loo, J.E. Tanner and M.R. Deveney (2010), "Toward routine, DNA-based detection methods for marine pests", *Biotechnology Advances* 28:706-714.

Bragg, J.R., R.C. Prince, E.J. Harner and R.M. Atlas (1994), "Effectiveness of bioremediation for the *Exxon Valdez* oil spill", *Nature* 368:413-418.

Browdy, C.L., G. Hulata, Z. Liu Z. G.L. Allan, C. Sommerville, T. Passos de Andrade, R. Pereira, C. Yarish, M. Shpigel, T. Chopin, S. Robinson, Y. Avnimelech and A. Lovatelli (2012), "Novel and emerging technologies: can they contribute to improving aquaculture sustainability?", in R.P. Subasinghe et al. (eds). "Farming the Waters for People and Food", Proceedings of the Global Conference on Aquaculture 2010, Phuket, Thailand. 22–25 September 2010. pp. 149–191. FAO, Rome and NACA, Bangkok.

Brown, L.R. (2010), "Microbial enhanced oil recovery (MEOR)", *Current Opinion in Microbiology* 13(3):316-320.

Brennan, L. and P. Owende (2010). "Biofuels from microalgae. A review of technologies for production, processing, and extractions of biofuels and co-products". *Renewable Sustainable Energy Reviews* 14: 557–577

Cabello, F.C. (2006), "Heavy use of prophylactic antibiotics in aquaculture: a growing problem for human and animal health and for the environment", *Environmental Microbiology* 8:1137–1144.

Chin Y.W., M.J. Balunas, H.B. Chai and A.D. Kinghon (2006), "Drug discovery from natural sources", *AAPS Journal* 8(2)e:239-253.

Chopin, T. (2010), *Integrated Multi-Trophic Aquaculture, Advancing the Aquaculture Agenda*, Workshop Proceedings, OECD Publishing, Paris. doi: http://dx.doi.org/10.1787/9789264088726-en.

Christensen, L. and R. Sims (2011), "Production and harvesting of microalgae for wastewater treatment, biofuels, and bioproducts", *Biotechnology Advances* 29:686–702.

Costanza, R., R. d'Arge, R. de Groot, S. Farberk, M. Grasso, B. Hannon, K. Limburg, S. Naeem, R.V. O'Neill, J. Paruelo, R.G. Raskin, P. Suttonkk and M. van den Belt (1997), "The value of the world's ecosystem services and natural capital", *Nature*, 387, 15 May.

Coyle, W.T. (2010), "Next-Generation Biofuels: Near-Term Challenges and Implications for Agriculture No. BIO-01-01, United States Department of Agriculture Economic Research Service, May. www.ers.usda.gov/media/122837/bio0101.pdf.

Cunningham, C.O. (2002), "Molecular diagnosis of fish and shellfish diseases: present status and potential use in disease control", *Aquaculture* 206(1-2):19-55.

Daar, A.S., P.A. Singer, D.L. Persad, S.K. Pramming, D.R. Matthews, R. Beaglehole et al. (2007), "Grand challenges in chronic non-communicable diseases", *Nature* 450:494–496.

Day, J.G., P.S. Slocombe and M.S. Stanley (2012), "Overcoming biological constraints to enable the exploitation of microalgae for biofuels", *Bioresource Technology* 109:245-251.

Dwyer, D.J., M.A. Kohanski and J.J. Collins (2009), "Role of reactive oxygen species in antibiotic action and resistance", *Current Opinion in Microbiology* 12:482–489.

FAO (2011). How to Feed the World in 2050: Issues Briefs, High-Level Expert Forum, Rome, 12-13 October 2009. www.fao.org/fileadmin/templates/wsfs/docs/expert_paper/How_to_Feed_the_World_in_2050.pdf

FAO/NACA (2012), "Farming the Waters for People and Food", R.P.Subasinghe, J.R. Arthur, D.M. Bartley, S.S. De Silva, M. Halwart, N. Hishamunda, C.V. Mohan and P. Sorgeloos (eds.), Proceedings of the Global Conference on Aquaculture 2010, Phuket, Thailand, 22-25 September, FAO, Rome and NACA, Bangkok.

Girguis, P.R., M.E. Nielsen and I. Figueroa (2010), "Harnessing energy from marine productivity using bioelectrochemical systems", *Current Opinion in Biotechnology* 21:252-258.

Gokulkrishnan, K., S. Kusuma and K. Boopalan (2011), "Antimicrobial activity of marine bacteria isolated from the Mangalore coast, west coast of India", *Recent Research in Science and Technology* 3(4):15-17.

Ho, S.H., C.Y. Chen, D.J. Lee and J.S. Chang (2011), "Perspectives on microalgal CO2-emission mitigation systems – A review", *Biotechnology Advances* 29:1189-198.

Hu G.P., Yuan J., Sun L., She Z.G., Wu J.H., Lan X.J., Zhu X., Lin Y.C. and Chen S.P. (2011), "Statistical Research on Marine Natural Products Based on Data Obtained between 1985 and 2008", *Marine Drugs* 9(4): 514–525, doi: http://dx.doi.org/10.3390/md9040514.

Jayakumar, R., M. Prabaharan, S.V. Nair and H. Tamura (2010), "Novel chitin and chitosan nanofibres in biomedical applications", *Biotechnology Advances* 28:142-150.

Leopoldina (2012), "Bioenergy – chances and limits", Deutsche Akademie der Naturforscher Leopoldina, Halle German National Academy of Sciences.

Mayer, A.M.S., K.B. Glaser, C. Cuevas et al. (2010), "The odyssey of marine pharmaceuticals: a current pipeline perspective", *Trends in Pharmacological Sciences* Vol. 31: 6.

Mussatto, S.I., G. Dragone, P.M.R. Guimarães, J.P.A. Silva, L.M. Carneiro, I.C. Roberto, A. Vicente, L. Domingues and J.A. Teixeira (2010), "Technological Trends, global market, and challenges of bio-ethanol production", *Biotechnology Advances* 28:817-830.

Newman, D.J. and G.M. Cragg (2007), "Natural products as sources of new drugs over the last 25 years", *Journal of Natural Products* 70(3):461-464.

Nutreco (2010). "Can we feed 9 billion people in 2050 in a sustainable way?" Nutreco, Amersfoort, Netherlands.

OECD (2008), *Biofuel Support Policies: An Economic Assessment*, OECD Publishing.
doi: http://dx.doi.org/10.1787/9789264050112-en.

Potters, G., D. Van Goethem and F. Schutte (2010), "Promising Biofuel Resources: Lignocellulose and Algae", *Nature Education* 3(9):14.

Reimers, C.E., L.M.Tender, S. Fertig and W. Wang (2001). "Harvesting energy from the marine sediment-water interface", *Environmental Science and Technology* 25:192-195.

Satpute, S.K., I.M. Banat, P.K. Dhakephalkar, A.G. Banpurkar and B.A. Chopade (2010). "Biosurfactants, bioemulsifiers and exopolysaccharides from marine microorganisms", *Biotechnology Advances* 28:436-450.

Scott, S.A., M.P. Davey, J.S. Denis, I. Horst, C.J. Howe, D.J. Lea-Smith and A.G. Smith (2010). "Biodiesel from algae: challenges and prospects", *Current Opinion in Biotechnology* 21:277-286.

Sommerset, I., B. Krossoy, E. Biering and P. Frost (2005), "Vaccines for fish in aquaculture", *Expert Review of Vaccines* 4(1):89-101.

Spier, C.R., G.G. Vadas, S.L. Kaattari and M.A. Unger (2011). "Near real-time, on-site, quantitative analysis of PAHs in the aqueous environment using an antibody-based biosensor", *Environmental Toxicology and Chemistry* 30(7) 1557–1563.

Sustainable Energy Authority of Ireland (SEAI) (2009). "A review of the potential of marine algae as a source of biofuel in Ireland", Sustainable Energy Authority of Ireland (SEAI), Dublin, February.

Tan T., Yu J. and Shang F. (2011). "2.58 - Biorefinery Engineering" in *Comprehensive Biotechnology* (Second Edition), Vol. 2:815-828.

Trincone, A. (2011), "Marine Biocatalysts: Enzymatic Features and Applications", *Marine Drugs* 9(4): 478–499.

Varmus, H., R. Klausner, E. Zerhouni, T. Acharya, A.S. Daar and P.A. Singer (2003), "Grand Challenges in Global Health", *H. Science* 302(5644):398-399, 17 October.

Wargacki, A.J., E. Leonard, M.N. Win, D.D. Regitsky, C.N.S. Santos, P.B. Kim, S.R. Cooper, R.M. Raisner, A. Herman, A.B. Sivitz, A. Lakshmanaswamy, Y. Kashiyama, D. Baker and Y. Yoshikuni (2012), "An Engineered Microbial Platform for Direct Biofuel Production from Brown Macroalgae", *Science* 335 (6066):308-313, 20 January.

Waters, A.L., R.T. Hill, A.R. Placeand M.T. Hamann (2010), "The expanding role of marine microbes in pharmaceutical development", *Current Opinion in Biotechnology* 21:780-786.

World Health Organization (2012). Grand Challenges in Genomics for Public Health in Developing Countries: Top 10 policy and research priorities to harness genomics for the greatest public health problems, www.who.int/rpc/grand_challenges.pdf, accessed July 2012.

Zheng, M.H., K. Hinterkeuser, K, Solomon, V. Kunert, N.J. Pavlos and Xu J. (2007), "Collagen-derived biomaterials in bone and cartilage repair", *Macromolecular Symposia* 253:179-185.

Chapter 3

Contributing to the bioeconomy: The economic potential of marine biotechnology

This chapter discusses existing and potential markets and other economic benefits to be realised through marine biotechnology. In the current economic situation, understanding the impact of investments and maximising returns on investment are more important than ever. This chapter therefore looks at existing measures and indicators for marine biotechnology and considers what further measures and indicators may be required.

Renewed interest in marine biotechnology has arisen in parallel with the birth and growth of the notion of the bioeconomy, the economic sectors that are based on bioscience and biotechnology innovation.[1] In 2009, the OECD's International Futures Programme undertook a project, The Bioeconomy to 2030, which examined the ways in which the bioeconomy is likely to create economic and social benefits for OECD and non-OECD countries.

Since that time, the term "bioeconomy" has become firmly entrenched in the lexicon of most OECD countries and has been taken up with different degrees of urgency. In 2009, Finland included a specific national bioeconomy strategy in its Council of State Natural Resources Strategy[2] and other Europeans countries (Denmark, Germany, Ireland and the Netherlands) also have national bioeconomy strategies. More recently, in February 2012, the European Commission announced a vision for the European Bioeconomy (European Commission, 2012). It estimated that the EU's bioeconomy sectors are worth EUR 2 trillion in annual turnover and account for more than 22 million jobs and approximately 9% of the workforce.

Outside Europe, Canada, the People's Republic of China and South Africa[3] either have or are planning their own ambitious strategies, and in April 2012 the United States published the US National Bioeconomy Blueprint.[4] This document recognises the bioeconomy as a political priority because of its tremendous potential for economic growth and social benefits. It considers that the bioeconomy will allow US citizens to live longer, healthier lives, reduce national dependence on oil, address key environmental challenges, transform manufacturing processes, and increase the productivity and scope of the agricultural sector while creating new jobs and industries.

Most of these bioeconomy strategies or visions include references to marine bioresources or marine biotechnology, and many of the opportunities identified and discussed in these documents and in the OECD's *Bioeconomy to 2030* (OECD, 2009a) have parallels in the field of marine biotechnology. For instance, food production and biofuels (agricultural biotechnology), development of new drugs (health biotechnology), of new materials (industrial biotechnology) and of bioremediation technologies (environmental biotechnology) are all potential sectors of application of marine biotechnology.

This indicates that marine biotechnology can make an important contribution to the bioeconomy through the development of innovative products and processes, the creation of jobs and the building or "greening" of a number of industries and sectors.

The market value of marine biotechnology

The global market for marine biotechnology products and processes is believed to offer a significant and growing economic opportunity. The widely cited Global Industry Analysts, Inc., estimated the global market for marine biotechnology at EUR 2.8 billion (2010 estimate), with a compound annual growth rate (CAGR) of 4-5% (or 10-12% under less conservative assumptions).[5] However the value of its contribution is difficult to quantify given the wide range of marine biotechnology applications and the difficulty of measuring and tracking these different markets. Marine biotechnology appears to differ from other biotechnologies in that it is defined in terms of its source (or target) material rather than the market it serves.

Nevertheless it is useful, and in many cases necessary, to quantify the market and market potential of marine biotechnology in order to attract investment and to guide policy development. However, the paucity of data and the fragmented markets only allow for estimating the value of a few markets. While this is a laborious and somewhat imprecise approach, it may help to illustrate the market potential of the field.

Pharmaceutical products

It is possible to quantify the market for some marine-sourced drugs and bioactive compounds because the industry is well established and data on commercialised products are publicly available. An early market success of marine biotechnology involved the extraction of the arabinosides Ara-A (Vidarabine®, Vidarabin®, Thilo®) and Ara-C (Cytarabine, Alexan®, Udicil®) from a sponge, *Cryptotethya crypta*, in the 1950s (Bergmann and Feeney, 1951). These compounds have anti-viral (Ara-A) and anti-leukemic (Ara-C) properties and an annual market of USD 50-100 million.[6]

The anti-inflammatory and analgesic pseudopterosins isolated from a Bahamian soft coral, *Pseudoterigorgia elisabethae*, are another useful example. These marine-derived compounds led to the development of bioproducts now used in skin care and cosmetics lines and are currently worth USD 3-4 million a year.[7]

Biotechnology

Several well-characterised bioactive compounds, such as shrimp alkaline phosphatase (SAP), isolated from organisms living in cold aquatic environments, and a thermostable DNA polymerase enzyme, isolated from the thermophilic bacterium *Thermus aquacticus* in hot springs, have found considerable commercial success in the biotechnology sector.

In 1985, Kary Mullis described how this thermostable DNA polymerase, known as *Taq* polymerase, could be used to amplify *in vitro* targeted DNA or RNA sequences rapidly using the polymerase chain reaction (PCR).[8] The Cetus Corporation patented the enzyme and associated technique and sold it to Hoffman-LaRoche for USD 300 million in 1991. PCR using *Taq* polymerase and other synthetic polymerases with similar properties is now used in biotechnology laboratories worldwide and represents a considerable market: sales of *Taq* DNA polymerase in Europe alone were USD 26 million in 1991 (Roberts, 1992) and had an initial estimated annual market of USD 50-100 million. The market for DNA polymerases is now believed to be in the order of USD 500 million a year.[9]

Fish and shellfish

The aquaculture industry is another market for which data exist. It is, however, difficult to assess the contribution of marine biotechnology to total market value. Taking commercial salmon production[10] as an example, the worldwide production of farmed Atlantic salmon exceeded 1.4 million tonnes in 2009 for a market value of USD 6.4 billion[11] (FAO, 2011). However, even though molecular aquaculture[12] is a significant part of salmonid aquaculture in most regions, the contribution of marine biotechnology to production and market value is not known. Also not captured in this market evaluation is the value of marine biotechnology used during production for the genotyping of eggs associated with selective breeding practices, PCR-based screening for fish health, or vaccine development.

Biomass-related markets

Markets for biomass-derived products, many of them marine-derived, are generally well established and offer some useful data. Seaweed-derived polysaccharides (including those derived from agar, alginates and carrageenan) have mature and relatively stable markets: 86 000 tonnes and USD 1 018 billion in 2010 (Bixler and Porse, 2010). These marine-derived compounds and derivatives are used in sectors ranging from food supplements to cosmetics to health care. In 2005, it was estimated that the biopolymer "woundcare" sector in which alginates and chitin are found was worth USD 800 million a year (Anonymous, 2005).

Globally the markets for chitin and chitosan (both largely marine-derived) are worth USD 481 million and are dwarfed by the market for chitin and chitosan derivatives (e.g. glucosamine) which are forecast to reach USD 63 billion and USD 21.4 billion, respectively, by 2015.[13] However, beyond specific products, it is not easy to break down the contribution of marine biotechnology to these global markets.

The market for algal biomass (for use in biofuels) is small and immature but is expected to grow exponentially in the next 5-10 years as demand increases. The size and value of the market has yet to be determined and will clearly depend on externalities such as production costs and rates, life-cycle analysis and government policies regarding use of renewable fuels.

The market for functional foods and natural products, including dietary supplements, natural and organic foods and beverages, functional foods and beverages, and natural and organic personal care and household products, was estimated at USD 270 billion in 2008[14] and is forecast to grow at around 6% through 2015. Again, with a few notable exceptions, it is difficult to separate out the fraction derived from marine bioresources. The global market for marine and algae oil omega-3 ingredients, estimated at USD 244 million in 2009, is forecast to reach USD 476-664 million by 2015 (based on estimated annual growth rates of 10.9% to 17.3%).[15]

Economic contribution of the oceans

Marine biotechnology contributes to the bioeconomy by creating jobs throughout the value chain from academic positions to positions in industry. However, it is also expected to affect many value-added sectors: pharmaceuticals, food, industrial processing, nutraceuticals, etc. This makes its precise economic impact difficult to determine. Some useful information might be obtained by looking at the economic contribution of the oceans as a whole, but data from existing studies suggest that marine biotechnology accounts for a relatively small fraction of marine-related activities, which include oil and gas, tourism, ship building, shipping, ports, etc.

In the United Kingdom in 2005-06, direct marine-related activities accounted for 4.2% (GBP 46 billion, at base prices) of total UK gross domestic product (GDP), and marine-related jobs represented 2.9% (890 000) of total UK employment. This equates to a total direct and indirect contribution to the UK economy of between 6.0% and 6.8%.[16] Marine-related R&D accounted for less than 1% of these economic activities (Pugh, 2008).

In Canada it was estimated that in 2001 marine activities contributed 1.4% to Canadian GDP although in the maritime provinces of British Columbia and Nova Scotia the percentages were 7% and 10%, respectively (Pugh, 2008). More recently, a Canadian report on the economic impact of marine activities in large ocean management areas (Pinfold, 2009) estimated that marine activities contributed 16.1% to GDP and accounted for 127 000 jobs.

In the United States, the National Ocean Economics Program (NOEP) estimated that in 2009 the ocean economy of coastal states represented value of USD 223 billion and accounted for 2.6 million jobs. The economic value derived from "living resources" (fish hatcheries and aquaculture, fishing, seafood markets and seafood processing) in the United States was estimated for 2009 at USD 5.7 billion and 58 000 jobs.[17] However, it is difficult to separate out the contribution of marine biotechnology to these economic statistics.

Despite some very tangible market successes, and some accurate regional economic data, quantifying the contribution of marine biotechnology to the bioeconomy remains a significant challenge. However, in today's world, determining the impact of investments and maximising return on investment is more important than ever.

Measuring inputs to marine biotechnology

Some believe that the potential of marine biotechnology is equal to that of land-based biotechnology, but that the field is too young to be measured by economic output indicators and should be measured using R&D and innovation indicators.

The OECD Scientific and Technological Indicators Database lists a number of input indicators related to R&D, such as gross domestic expenditure on research and experimental development (GERD) and financing patterns, to measure the output of scientific and technological activities. In particular, it contains three proxy indicators for innovation that could be useful in this regard: patents, the technology balance of payments and trade in R&D-intensive industries (OECD, 2009b). Most consideration to date has been given to patent data.

Patent data

Patents are often a significant part of the overall process of commercialising innovations. This is the case for marine–based resources and related innovations and, for this reason, patents can to some extent be considered a proxy for the commercial value of discoveries (Leary et al., 2009). Some data on marine biotechnology-related patents are available and worth examining as an indication of the growth and economic potential of the field.

In one survey, patenting was studied over a ten-year period using the database search terms: marine, sea, ocean, deep water, and sea water. This study identified a total of 2 241 patents related to marine biotechnology granted during 1996-2005 in nine different fields (Anonymous, 2010).

Figure 3.1. Marine genetic resource patents (patents files, n=135)

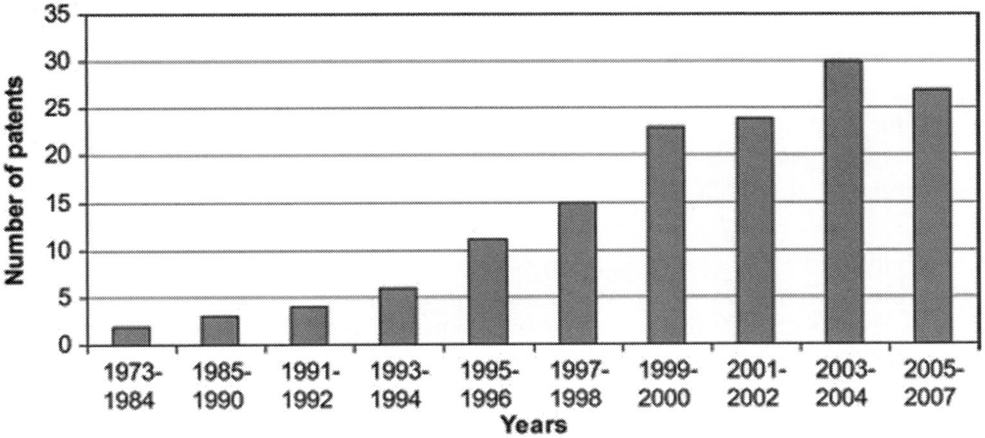

Other sources (in Hunt and Vincent, 2006): 14 000 novel chemicals; 300 patents on marine natural products.

Source: Adapted from Leary et al. (2009), "Marine genetic resources: A review of scientific and commercial interest", *Marine Policy*, 33:183-194.

Figure 3.2. Patent distribution

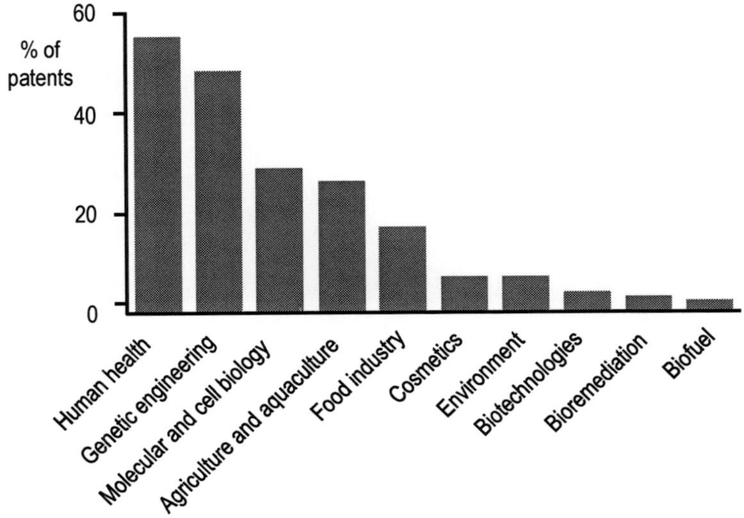

Source: Arrieta et al. (2010), "What lies underneath: Conserving the oceans' genetic resources", *Proceedings of the National Academy of Sciences* 107(43):18300-18305.

Leary et al. (2009) looked at patents associated with marine genetic resources (Figure 3.1) and identified 135 patents filed between 1973 and 2007. Arrieta et al. (2010) also restricted their analysis of patents to those associated with the genes of marine organisms and identified 460 claims in ten different fields of use (Figure 3.2). The available data indicate an increase in the number of marine biotechnology patents associated with marine genetic resources and demonstrate the wide variety of applications associated with patents in just one area of marine biotechnology.

However, in this field, as in many others, the use of patents as an output indicator of R&D or as a proxy for commercial success is limited. Not all inventions patented are commercialised and not all commercial successes are patented. Additionally, differences in patenting practices between publicly funded and privately funded research complicate the picture, since patenting is more prevalent and happens at a faster rate in the private sector (Leary et al., 2009).

Other indicators

The Marine Biotechnology Working Group of the Marine Board (European Science Foundation) attempted to map indicators of success and found that it was very difficult, and in some cases impossible, to obtain the necessary information. The working group was able to measure some key parameters of scientific outputs: funds and manpower devoted to marine research and technological development; scientific publications and their impact (citations); European patents by marine science and technology sectors; and information on the objectives, current status and results of various research and technological development initiatives and programmes, both at national and European level (ESF, 2002). However, information pertaining to businesses and economic outputs was hard or expensive to obtain and difficult to interpret.

The need for new measures

Measuring the commercial successes and economic impact of investment in science and technology is important (OECD, 2010). It is especially important for emerging fields such as marine biotechnology which stand to be significantly influenced by policy. There is a need to assess the impact of direct and indirect investment by governments and other stakeholders and to measure progress along the discovery and development continuum.

For marine biotechnology, some information can clearly be gained by looking at some of the indicators mentioned above. However, these are generally inadequate indicators of eventual commercial success. This raises the question of how governments and other stakeholders can monitor and assess the development of the field and its contribution to the economy. It may be that new measures and indicators are needed, or that new data need to be collected.

In some sectors indicators and measures are already in place. For example, a large body of statistics relating to R&D inputs and outputs is available for the pharmaceutical industry. Moreover, this industry is actively exploring the potential of marine-derived compounds to help strengthen a drug development pipeline that has suffered from declining productivity in recent years (Schmid and Smith, 2005; Di Masi et al., 2003; Peck, 2007).[18] In this sector, the rise in marine-related patents as a proportion of all patents related to drug development can be used as an indicator of the increasing contribution of marine biotechnology. This output, linked to marine biotechnology R&D inputs such as funding, may provide a partial picture of the economic impact of marine biotechnology. However a more complete picture will be required both for this sector and for others.

Development of appropriate measures and indicators of inputs and economic outputs will require a common definition of marine biotechnology. This would make it possible to see the types of investments being made and the types of innovations being produced. The work might also include an analysis of the business models used by certain sectors and a broad range of socioeconomic indicators in order to describe the status and evolution of marine-related activities, such as economic added value and the employment generated by various branches of marine research and technology. For reasons of manageability, the focus could initially be on a few countries and then be extended and validated by a study of additional countries.

Non-market value of the ocean

In striving to measure, and realise, the economic potential of the ocean and its bioresources, it is important to consider the non-market value of the oceans, e.g. the environmental (ecosystem services) and recreational value that can be derived from the ocean, and to recognise how these are affected, positively and negatively, by marine biotechnology applications.

In a controversial and widely cited study, Costanza et al. (1997) estimated the economic value of 17 ecosystem services for 16 biomes using a number of published studies and original calculations. They looked at the value of gas regulation, food production, waste treatment, climate regulation, etc., and estimated that the average global value of ecosystem services provided by the marine environment (open ocean, coastal regions) to be USD 20 949 x 10^9 a year.

The National Ocean Economic Program (NOEP) at the Center for the Blue Economy at the Monterey Institute of International Studies has also looked at the non-market value of oceans. Its website[19] provides links to a number of (mainly US) studies related to the non-market value of oceans used in their valuation.

Non-market values are often linked to the recreational benefits of ocean and coastal environments, or the environmental services they supply, but these values also extend beyond any direct-use benefits that the oceans and coasts provide. NOEP tries to estimate the value to society of things such as pristine beaches in California, abundant wildlife in the Florida Keys, or wetland and mangrove systems that help mitigate storm damage off the Gulf Coast (Anonymous, 2006). While the value of these "intangibles" is difficult to estimate, it is important to try to include their value when measuring the contribution of the marine environment to the economy.

Non-market values are not insignificant. In the United States alone, the NOEP analysis suggests that the total non-market value of ocean and coastal resources is, at a minimum, tens of billions of dollars a year and likely to be much more. In Florida, for example, it is estimated that the non-market value of seven activities ranged from approximately USD 16.5 billion to USD 53 billion a year (Pendleton, 2007). Excluding these non-market values would underestimate the true value of the ocean economy.

An alternative to considering the non-market value of the ocean and marine resources is to consider market values in the absence of these resources. The Stockholm Environment Institute takes this approach and looks at the lost value of the oceans under different climate change scenarios (Table 3.1). Even in a low climate impact scenario, it is estimated that over USD 1 trillion in value (0.06% of GDP) may be lost by 2050.

Any economic assessment of ocean and coastal resources should include a thorough accounting of market and non-market values to enable well-informed decisions regarding the use and development of ocean resources. Given the difficulty of valuing these less tangible or quantifiable non-market values, and the delicate balance to be achieved between ocean productivity and sustainability, a need for additional socioeconomic and environmental indicators of ocean health has been suggested (ESF, 2002).

Table 3.1. Valuation of selected climate impacts on the ocean (USD billions of 2010)

	Low climate impacts		High climate impacts		Difference	
	2050	2100	2050	2100	2050	2100
Fisheries	67.5	262.1	88.4	343.3	20.9	81.2
Sea-level rise	10.3	34.0	111.6	367.2	101.3	333.2
Storms	0.6	14.5	7.0	171.9	6.4	157.4
Tourism	27.3	301.6	58.3	639.4	31.1	337.7
Ocean carbon sink	0.0	0.0	162.8	457.8	162.8	457.8
Total	105.7	612.2	428.1	1 979.6	322.5	1 367.4
Percent of GDP	0.06%	0.11%	0.25%	0.37%	0.18%	0.25%

Source: Valuing the Ocean: Draft Executive Summary, Stockholm Environment Institute, www.sei-international.org/publications?pid=2064.

Indicators of healthy oceans

Environmental indicators could contribute to effective resource management and protection protocols. These might include biological, geological, chemical and physical indicators that characterise the health of coastal waters, the nature of pollutants and their relation to human activities and urban concentration. While some national indicators and information may exist, they are generally limited and not comparable among countries (ESF, 2002). Further work is required to:

- Define and analyse the policy value of relevant quantitative indicators.
- Identify existing primary science and technology indicators and socio-economic data on a sectoral and national basis.
- Analyse the validity and relevance of such indicators and data for policy development, such as a demonstration of sustainable development options adapted to regions.
- Synthesise existing indicators with a view to developing international indicators, including benchmarking of indicators and practice.
- Publish and disseminate regular reports on the state of the ocean and on marine activities based on these indicators.

Such data could contribute to comprehensive databases on scientific, technical and socioeconomic competencies relevant to policy making.

Conclusion

Marine biotechnology can contribute to the bioeconomy through the development of innovative products and services in sectors such as food, health and manufacturing and through job creation. To the extent that marine biotechnology can contribute to the sustainable use of ocean bioresources, it can help to preserve the non-market value of the ocean and associated socio-economic benefits (e.g. recreation, cultural traditions, tourism). The ability to measure the socioeconomic contribution of marine biotechnology is important for a number of reasons and will necessarily underpin and influence its future development.

The market value of some marine biotechnology products and services is known, yet for others the size and value of the market is difficult to estimate. Difficulties arise both for tracking the range of products and services across different sectors and for separating out the contribution of marine biotechnology from other factors. It will be necessary to reach a common understanding or definition of marine biotechnology in order to develop appropriate indicators of inputs and outputs. Given the range of applications of marine biotechnology, development of indicators and measures might usefully focus initially on a few products or outputs in a few countries before being extended to and validated for marine biotechnology outputs in other countries and sectors. The larger goal is the development of economic indicators and metrics suitable for comparative analysis across countries and over time.

Owing to the delicate balance to be struck between ocean productivity and sustainability there is a need for indicators that can provide an "economic assessment" of healthy ecosystems. These indicators might include measures of biodiversity and pollution and provide information regarding the fitness of ocean bioresources, as these are the foundation of marine biotechnology and thus of its economic potential.

Notes

1. As a relatively new concept the term "bioeconomy" is interpreted in different ways by different actors (OECD, 2009).
2. www.sitra.fi/en/natural-resources-strategy, accessed August 2012.
3. www.pmg.org.za/report/20120222-department-science-technology-grand-challenges-bioeconomy-committee-d, accessed August 2012.
4. www.whitehouse.gov/sites/default/files/microsites/ostp/national_bioeconomy_blueprint_april_2012.pdf, accessed August 2012.
5. www.prweb.com/releases/2011/1/prweb8041141.htm, accessed March 2011.
6. http://aquafind.com/articles/Marine-Biotechnology.php.
7. Mayekar et al., n.d., "Marine biotechnology: Bioactive natural products and their applications", http://aquafind.com/articles/Marine-Biotechnology.php.
8. DNA polymerases have been isolated from hyperthermophiles for use in the process of DNA amplification known as polymerase chain reaction. These unusually thermostable polymerase enzymes, such as *Taq* DNA polymerase, must withstand the alternating cycles of heating and cooling in the PCR process if the target DNA is to replicate. See: http://nano.nstl.gov.cn/sea/MirrorResources/1895/Extremophiles.cfm.html.
9. www.highseasconservation.org/documents/juniper.pdf.
10. Salmonid aquaculture has been growing steadily since the late 1970s. It is the source of the majority of salmon available in most markets and may become an environmentally sustainable response to the depletion of wild stocks by capture fisheries.
11. www.fao.org/fishery/culturedspecies/Salmo_salar/en.
12. Molecular aquaculture involves the application of molecular techniques to rearing of fish through genetic breeding, molecular fish health diagnostics, vaccines, etc. See http://ec.europa.eu/research/bioeconomy/pdf/cwg-mb_to_kbbenet_report_final.pdf.
13. www.prweb.com/releases/chitin_chitosan/derivatives_glucosamine/prweb4603394.htm, accessed August 2012.

14. http://newhope360.com/nutrition-business-journal-reviews-270-billion-global-nutrition-industry, accessed August 2012.

15. www.frost.com/prod/servlet/press-release.pag?docid=207388177, accessed August 2012.

16. The authors write: "The scope is taken to include those activities which involve working on or in the sea. Also those activities that are involved in the production of goods or the provision of services that will themselves directly contribute to activities on or in the sea. This restricted definition is based on the understanding that the figures produced are minimum estimates of the economic importance of marine resources and activities."

17. www.oceaneconomics.org/Market/ocean/oceanEcon.asp.

18. In addition to rising costs and traditionally long development times, new drugs represent a very small (7% in 2009) and decreasing portion of sales for the industry as a whole, with many companies including in their portfolio generic drugs and easily developed "me-too" compounds developed by other companies targeting the same disease and relying on a similar action mechanism, www.reuters.com/article/2010/06/27/us-pharmaceuticals-rd-idUSTRE65Q3IM20100627, accessed January 2012.

19. www.oceaneconomics.org/nonmarket/

References

Anonymous (2005), Prospects for Marine Biotechnology in the UK, Vol. 2.

Anonymous (2006), Florida's Ocean and Coastal Economies Report, NOEP (National Ocean Economics Programme), Washington DC.

Anonymous (2010), Master Plan for Marine Biotechnology: Blue-bio 2016. Korean Ministry of Land, Transport and Maritime Affairs.

Arrieta, J.M., S. Arnaud-Haond and C.M. Duarte (2010), "What lies underneath: Conserving the oceans' genetic resources", *Proceedings of the National Academy of Sciences* 107(43):18300-18305.

Bixler, H.J. and H. Porse (2010), "A decade of change in the seaweed hydrocolloids industry", *Journal of Applied Psycology.* doi: http://dx.doi.org/10.1007/s10811-010-9529-3.

Costanza, R., R. d'Arge, R. de Groot, S. Farberk, M. Grasso, B. Hannon, K. Limburg, S. Naeem, R.V. O'Neill, J. Paruelo, R.G. Raskin, P. Suttonkk and M. van den Belt (1997), "The value of the world's ecosystem services and natural capital", *Nature*, 387, 15 May.

Di Masi, J.A., R.W. Hansen and H.G. Grabowski (2003), "The price of innovation: new estimates of drug development costs", *Journal of Health Economics* 22:151-185.

European Commission (2012). Innovating for Sustainable Growth: A Bioeconomy for Europe http://ec.europa.eu/research/bioeconomy/pdf/201202_innovating_sustainable_growth_en.pdf, accessed August 2012.

ESF (European Science Foundation-Marine Board) (2002). Integrating Marine Science in Europe. ESF Marine Board Position Paper 5, November 2002. www.vliz.be/imisdocs/publications/113786.pdf, Accessed August 2012.

FAO (2011). How to Feed the World in 2050: Issues Briefs, High-Level Expert Forum, Rome, 12-13 October 2009 www.fao.org/fileadmin/templates/wsfs/docs/expert_paper/How_to_Feed_the_World_in_2050.pdf.

Hunt, B. and A.C.J. Vincent (2006), "Scale and Sustainability of Marine Bioprospecting for Pharmaceuticals", *AMBIO: A Journal of the Human Environment* 2006 35 (2), 57-64

Leary D., M. Vierros, G. Hamon, S. Arico and C. Monagle (2009), "Marine genetic resources: A review of scientific and commercial interest", *Marine Policy*, 33:183-194.

OECD (2009a), *The Bioeconomy to 2030: Designing a Policy Agenda*, OECD Publishing.
doi: http://dx.doi.org/10.1787/9789264056886-en.

OECD (2009b). "Main Science and Technology Indicators", *OECD Science, Technology and R&D Statistics* (database).
doi: http://dx.doi.org/10.1787/data-00182-en.

OECD (2010), *The OECD Innovation Strategy: Getting a Head Start on Tomorrow*, OECD Publishing.
doi: http://dx.doi.org/10.1787/9789264083479-en

Peck, R.W. (2007), "Driving earlier clinical attrition: if you want to find the needle, burn down the haystack. Considerations for biomarker development", *Drug Discovery Today* 12:289-294.

Pendleton, L. (2007), "The Economic Value of Coastal and Estuary Recreation", in L. Pendleton (ed.), *The Economics of Estuary and Coastal Restoration: What's at Stake, Restore America's Estuaries.*

Pinfold, G. (2009), "Economic Impact of Marine Activities in Large Ocean Management Areas", *Statistical and Economic Analysis Series, Publication* No.1-2, Fisheries and Oceans Canada, Ottawa, Ontario.

Pugh, D. (2008). "Socio-economic Indicators of Marine-related Activities in the UK Economy", The Crown Estate, March, www.abdn.ac.uk/mrm/publication/socio_economic_uk_marine.pdf.

Roberts, L. (1992), "Roche gets tough on illicit sales of PCR reagent", *Science* 258:1572–1573.

Schmid, E.F. and D.A. Smith (2005), "Is declining innovation in the pharmaceutical industry a myth?", *Drug Discovery Today* 10:1031-1039.

Chapter 4

Infrastructures for generating and sharing knowledge

This chapter discusses the knowledge-based, scientific and technological infrastructure required to reap the benefits of marine biotechnology. In other fields, large national or international projects have drawn financial and political attention to the infrastructures required to meet their goals. The chapter asks: "What type of infrastructure is necessary to drive development of the field?" and "What policies might be required to achieve this goal?"

The extent to which the benefits of marine biotechnology are realised will depend, in large part, on how well marine bioresources and the marine ecosystems are understood and conserved. As the preceding chapter has argued, this will require the collection and analysis of new scientific data and comparison of the data with existing knowledge. Realising the full potential of marine biotechnology will also require appropriate research and development (R&D) infrastructures.

Research infrastructures

Heightened interest in marine biotechnology is linked to recent advances in "omic" technologies (e.g. genomics, proteomics) and the new insights into marine bioresources they have made possible. These technologies are fundamental to many marine biotechnology R&D activities and are producing a wealth of genetic data which can lead to better understanding of ocean life and its potential for biotechnological development. Genomic sequencing technology, once a technical and financial stumbling block, has matured and produces data at rates unheard of a decade ago. Illumina HiSeq technology, for example, produces some10 terabits (Tb) of sequence data per machine annually and output from future technologies is expected to increase to an annual 112 Tb per machine by 2015. Meanwhile, the cost per nucleotide of sequencing is dropping rapidly (Figure 4.1).

Figure 4.1. Progression of sequencing rates and cost

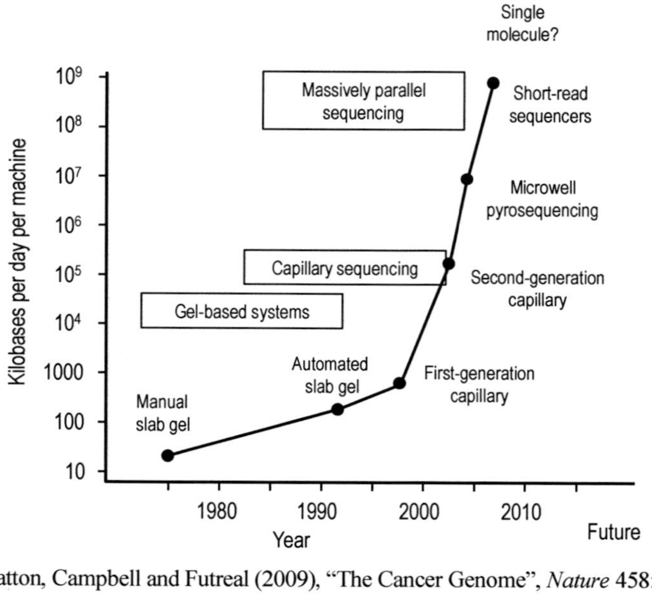

Source: Stratton, Campbell and Futreal (2009), "The Cancer Genome", *Nature* 458:719-724.

Despite the advances in "omics",[1] very little is known as yet about marine biodiversity and its potential in terms of bioresources. If, as some suggest (Mora et al., 2011; COML, 2010), something has been learnt about 10% of the ocean's species, this is undoubtedly the most accessible fraction, encompassing most of the larger mammals, fish and plants and the most common and easy-to-culture microbes. Of the rest – the largest share of marine organisms by number and weight – very little is known.

Metagenomics, or plurality sequencing, of marine microbial communities is opening an unprecedented window on biodiversity. The recent advent of single-cell sequencing (SCS) will further improve the capacity to link the structure and functions of microbial communities in the coming decade (Zhang et al., 2006; Su et al., 2012). Past investment in genomics has led to a marine biotechnology renaissance, which is leading to a new infrastructure bottleneck that threatens to limit the rate at which its benefits can be realised. The widespread application of marine biotechnology will clearly require additional infrastructure.

Tools and processes to collect, culture and catalogue samples

Tools and processes for collecting samples from the marine environment are improving all the time. However, exploration and sampling are still difficult in areas of environmental extremes which offer great potential for discovering organisms with novel functionalities. The Mariana Trench in the western Pacific is a good example. At its deepest point, it is over 10 km deep and the water column exerts a pressure of 1 086 bars (15 750 psi), over a thousand times the standard atmospheric pressure at sea level. As of mid-2012, four expeditions had been made to the bottom of the trench and several more are planned, but they are technically difficult, expensive and very risky. Sampling is limited and, although living organisms have been collected, culturing and studying them remains a challenge. In such extreme environments, technologically advanced vessels are needed for marine biotechnology R&D.

Culturing provides a means of undertaking in-depth analysis of biochemical networks and systems and of preserving marine resources in biobanks. The complex, symbiotic nature of marine organisms means that new culture methods will be required for basic research (e.g. model organisms, screening). In particular, new methods and media, perhaps developed with knowledge from metagenomic studies (and associated microbial communities), that enable cell-to-cell communication or signalling, may be required to culture as yet "unculturable" organisms and symbiotic organisms. New culturing methods based on co-metabolism between community members represent a radical change from the conventional "isolate and enrich" approach to cell culture.

At the production level, culturing may provide a solution to the problem of unsustainable harvesting and may be required for sustainable production of many new compounds/molecules/enzymes. New processes range from the optimisation of biorefineries to produce algal biofuels, to the culturing of genetically modified organisms (GMOs) for pharmaceuticals, to the culturing of bacteria and viruses in the laboratory, to the development of new cell lines.

Collections of biological specimens (or parts thereof) can be used to analyse and preserve biodiversity, to facilitate the exchange of resources and to develop model organisms. Biobanks containing living cultures, nucleic acid archives, extract libraries (such as those obtained with new chromatographic instruments and media) and compound libraries (which enable the study of structures, function and origin) can facilitate the development of new molecules, compounds and bioactives. Development of novel cryopreservetion techniques and capacity can also support biobanking efforts and commercial-scale culturing of larvae and marine organisms.

Databases

Databases are an integral part of the study of marine bioresources and biodiversity. A number of databases containing different types of information exist. The World Register of Marine Species (WORMS)[2], hosted at the Flanders Marine Institute, VLIZ, was established as a global effort to register the names of all marine species. The project involved 270 expert taxonomists from 185 institutions in 38 countries, and the database describes 215 000 species. WORMS is accessible on the Internet and is broken down into sub-portals for different taxonomic groups.

The UNESCO-IOC/IODE Ocean Biogeographic Information System (OBIS)[3] is the largest source of information on the distribution of marine species and is a data legacy of the ten-year Census of Marine Life (COML) programme. OBIS contains 32 million records from 1 000 datasets and over 100 000 marine species. Its geographical coverage is good for highly populated regions but less so for remote regions such as the open ocean, the deep sea and the polar regions. The database is an excellent resource for the study of marine organisms and biodiversity in the marine environment.

Both of these databases are publicly accessible and have an international base. However, to remain current, they will require access to networks of stations, seagoing platforms and observatories. They will also need to draw on a large team of taxonomists to ensure that the data are of high quality. This work offers opportunities for synergies. For instance, the establishment of marine observatories will provide an opportunity to collect excess material and bulk samples of water and ocean floor for preservation *in-situ* or *ex-situ* biobanks for later use and for as yet non-existent technologies. There is also a need to be able to couple the information in these databases with genetic, species and habitat information in other databases and to contextualise this information in terms of ecosystem parameters. Finally, the standardisation of protocols for sample collection and cataloguing will facilitate the assimilation of the work of other research groups, making for easier sharing of information

Platforms for screening and analysis

Genome sequence data (from whole genome data to metagenomics data) from the marine environment is certainly the main type of information used in marine biotechnology R&D. The data are produced rapidly and increasingly inexpensively and this is challenging current R&D infrastructures. For instance, the rate of sequence submissions to archival databases, which are fundamental for reproducible science, is outstripping the rate of growth of storage capacity (Kodama et al., 2012).[4]

Perhaps the most immediate challenge to marine biotechnology is the development of tools and platforms to facilitate high throughput screening of new "omics"-related information. Screening seeks to compare sequence data with information about known genes (including gene products and gene expression profiles) to infer the structure, function or identity of the sequence or organism of interest. These screens have historically relied primarily on comparison to annotated DNA sequences (and related information) in databases such as Genbank (Bensen et al., 2012) or the Marine Genomics Europe (MGE) Bioinformatics Portal.[5] However, given the relative lack of information about marine bioresources, and the speed at which new sequence data are being generated, these approaches are proving insufficient for accurate annotation[6] of genomic sequences from novel marine organisms. The existing infrastructure is also challenged by the biodiversity and complexity of the marine environment, which will require the development of new approaches and platforms to link genotype with phenotype from single cells to ecosystems.

Model systems, including *in silico* models (Lerman et al., 2012), have been developed for many organisms in an attempt to bridge the gap between genotype and phenotype. Model organisms provide a means of obtaining a better understanding of biochemical processes and thus of identifying pathways for modification (targets) for further development or production. Such models exist for a number of marine species of medical, industrial or commercial importance (e.g. salmon, sea urchin) or for evolutionary or developmental study (e.g. marine annelid worm).

Model systems support both basic research and the development of advanced marine biotechnology and thus facilitate a systems-based approach to genome annotation and elucidation of new gene functions. In the development phases, model systems can be used to access or harvest organisms or derivatives of interest. Synthetic biology may eventually find application in this field but until then, model organisms can help identify and refine the most cost-effective routes to harvesting or producing functional compounds or organisms of interest. Model systems will be especially useful for the study of new phyla or classes of organisms and extremophiles, which are considered a significant source of new functions. The definition of model systems will need to be expanded to include ecosystems and ocean observatories in order to evaluate community responses to environmental perturbation.

The concept of the "minimal genome" provides a useful way of linking genes to functions and has been useful for identifying minimal metabolic pathways and linking genes with function (Mushegian, 1999). For instance, Dufresne et al., (2003) published the sequence of *Prochlorococcus marinus*, one of the ocean's dominant photosynthetic organisms. It is one of the smallest photosynthetic organisms, and its genome approximates the minimal gene complement for a photosynthetic organism. Using both model organisms and minimal genomes, it is possible to identify and assign functions to unknown genes.

In recent years, a number of databases have been developed to facilitate the comparative analysis of species. These databases are the workhorse of marine biotechnology and contain molecular and genomic data from microbial communities and individual genomes. Like the Integrated Microbial Genomes (IMG) system (Markowitz et al., 2006, 2012), they serve as a community resource for comparative analysis and annotation of all publicly available genomes. They are free and publicly available.

The complexity of metagenomic data brings with it a further challenge for annotation linking genotype and phenotype. This challenge is concisely articulated by Chisholm and Cary (2001): "Our genetic and biochemical understanding of metabolism, and other cell functions, is based largely on the study of complete pathways within cells. However, microbial communities are a collection of gene functions distributed amongst its individual members which form distributed metabolic pathways directing matter and energy exchange among and between microbes. No single organism contains all the genes necessary to perform the diverse biogeochemical reactions that make up ecological community function." This realisation has important implications for how gene functions are defined or classified, within symbiotic or related groups of microorganisms, into meaningful units of selection, utilisation or conservation. The complexity of microbial communities is driving the development of a new wave of e-infrastructures (see Wright et al., 2012, for a review of the microbial ecology of expanding oxygen minimum zones).

Of particular importance will be the development of interactive services that allow for uploading user information for analysis and visualisation and enabling the study of comparative genomics and metagenomics. These e-infrastructures should take into account the multidimensionality of marine genomics data, which include physical and chemical properties and molecular information, in order to integrate metadata with sequence information in the taxonomic and metabolic pathway context.

The development of novel data management platforms and information services[7], and the generation of data products such as visual analytics and web services, are of paramount importance as growth opportunities for marine genomics and biotechnology. Developing the required infrastructures will be a significant undertaking, financially, structurally and operationally. It will involve considerations unique to marine bioresources and may therefore benefit from specific policy attention. Traditionally, national or multilateral collaboration on research infrastructures has been justified in terms of cost sharing or the need to generate economies of scale and scope. These considerations are even more important for the shared resources, and the associated data, of the very large and highly complex marine environment.

International partnerships to drive innovations in R&D infrastructure

The OECD Innovation Strategy recognises the impact that fully functioning knowledge networks can have on the efficiency and effectiveness of the innovation process, both stimulating innovation and improving its efficiency by reducing transaction costs (OECD, 2010). As marine biotechnology becomes a focus of investment and innovation strategies, it will be important to ensure that mechanisms are in place to generate, share and give value to knowledge in order to enable innovation. Given the global nature of marine bioresources, it will also be useful to consider international, trans-boundary approaches that can help to drive innovation in R&D infrastructure. The existence of several marine biotechnology funding programmes and initiatives suggests that the value of international partnerships and investment for fostering knowledge development is recognised.

EU Joint Programming Initiative: A regional co-ordinated approach to investment

Within the European Union (EU), the Joint Programming Initiative (JPI) provides one model for resource sharing and the co-creation of knowledge. The EU JPI grew from the recognition that "Europe not only needs to invest more in research, but also needs to invest it to better effect if it is to achieve its declared vision: a balanced and sustainable development,"[8]

The purpose of the JPI Oceans (www.jpi-oceans.edu)[9] is to increase the value of national R&D investments in ocean research in the Europe Union in order to avoid fragmentation and unnecessary duplication of R&D, to look for synergies, and to facilitate different types of co-operation to meet policy objectives and global challenges. The benefits of JPI Oceans include a long-term perspective (10-15 years), a high level of commitment and voluntary participation in different actions. JPI Oceans will focus on: the development of new bioactive products for health treatments; biotechnology applied to aquaculture; biofuel from marine algae; screening of marine genetic diversity; development of marine biosensors to monitor the environment; and mitigation of human and climate change impacts on the ocean. This initiative is an indication of renewed attention to international research partnerships based on shared goals, understanding and challenges.

The EU's ERA-NET (the European Research Area Networks scheme) was also designed to foster the co-ordination of national and regional research programmes. ERA-NET activities related to marine biotechnology under the EU's 7th Framework Programme (FP7) aim at transnational co-operation in this area.

Regional partnership with a regional focus

The much older Mediterranean Science Commission (CIESM), established in 1908, is an international partnership with a regional focus. The Commission has grown to 22 member states which support the work of several thousand marine researchers who seek to understand, monitor and protect the rapidly changing, highly stressed Mediterranean Sea. The CIESM takes a multidisciplinary approach. It encourages the exchange of scientific standards and ideas and maintains a constructive dialogue around the Mediterranean Sea and the Black Sea basins, so that the region benefits from collaboration among researchers and populations.

The European marine biological resource centre: A distributed infrastructure

The European marine biological resource centre (EMBRC) is a collaboration of 12 leading marine stations and the EMBL (European Molecular Biology Laboratory) which uses the latest technologies to study marine organisms (microbes, plants, animals). Through a network of distributed research infrastructure with state-of-the-art research and training facilities, EMBRC enables the scientific community at large, including universities and industry, to access marine organisms, aquaria facilities, and dedicated platforms for genomics, structural and functional biology, microscopy, and bioinformatics. Through its network, EMBRC aims to provide comprehensive support (including interdisciplinary training) for the intelligent and sustainable exploitation of marine resources.

The Red Sea Research Centre: International collaboration with a regional focus

The Red Sea Research Centre at KAUST is another example of international collaboration with a regional focus. The Centre, with collaborators from the United Kingdom, the United States, the Netherlands, Hong Kong (China) and the Arabian Peninsula, is working to develop a scientific basis for sustaining and conserving coral reef environments along the Red Sea coast of Saudi Arabia. The group takes a multidisciplinary approach to learning about coral reef ecosystems and their oceanographic context, including the physical, chemical, biological and geological environment and the stresses arising from natural as well as anthropogenic factors such as overfishing, pollution, coastal development and global climate change.

BioMarks: A midsize international collaboration

In contrast to the other partnerships described, BioMarks (Biodiversity of Marine Eukaryotes) is a relatively small international collaboration of eight EU research institutions in four countries. Its goal is to develop "metagenetics" or "metabarcoding" to facilitate the study of global biodiversity. The group is tackling a specific challenge: the development of biomarkers (metabarcodes) that describe the taxonomic, genomic and metabolic diversity of natural environments. The 30 members of this multidisciplinary team examine coastal marine protist biodiversity using massive rDNA sequencing integrated into a network of taxonomic expertise and comprehensive contextual phenotypic and environmental metadata. If successful, metabarcoding and metagenomic data could be used for the biomonitoring of sites affected by human activities, for bioprospecting, for understanding the functioning of the global ecosystem and for reconstructing past environmental changes. The work will also provide a foundation for the future annotation of genomic sequences and taxonomically controlled eukaryotic metagenomics.

As governments consider investment in and development of research infrastructures for marine biotechnology, it will be important to take account of the lessons learned from these and other initiatives. It will be useful to look at synergies that can be achieved across initiatives and how these synergies can be exploited to best effect. It will also be essential to consider ways for developed countries to share scientific and technological infrastructures with emerging economies and developing countries. If tackled in the right way, these infrastructures could transform the way and rate at which knowledge is generated and used.

Marine biotechnology megaprojects

In other scientific, technical or engineering fields, large national or international projects have focused financial and political attention on developing the necessary infrastructures. This approach has been critical to success in fields such as the Human Genome Project (HGP). More than two decades after it began, the HGP is often used to illustrate the benefits of internationally co-ordinated, or harmonised, development of infrastructures: reduced duplication, easier data sharing, larger and more cohesive infrastructures; etc. From 1990 to 2003, the HPG concentrated considerable investment (USD 3.8 billion from the US government alone) and infrastructure on efforts to determine the sequence of the chemical base pairs that make up DNA and to identify and map the approximately 20 000-25 000 genes of the human genome.

Completed in 2003, the HGP left a legacy of genomics and bioinformatics infrastructure which continues to be used and developed by the international research community in both the public and private sectors. The enduring nature of this legacy contrasts with some other major "big science" projects with a much more finite life span, such as the Superconducting Super Collider (USD 11 billion, estimated life of 30 years) and the Hubble Space Telescope (USD 1.5 billion, estimated life of 15–20 years).

The HGP was also transformational in that it illustrated the potential for biology to be an economic driver. The US federal government investment in the HGP, through to its completion in 2003 (USD 5.6 billion in 2010 terms), was crucial to the generation of an economic output of USD 796 billion, giving a return on investment to the US economy of 141 to 1 (Battelle, 2011). In 2010 alone, the genomics-enabled industry generated over USD 3.7 billion in federal taxes and USD 2.3 billion in state and local taxes. In other words, in one year the government received revenue nearly equal to the entire 13-year investment in the HGP. Also in 2010, human genome sequencing projects and associated research and industry activity directly and indirectly created 310 000 jobs.

A more recent example is the 1000 Genomes Project, an international research effort to establish a detailed catalogue of human genetic variation by sequencing the genomes of 1 000 individuals.[10] At its peak, this international megaproject sought to generate over 8.2 billion bases per day (the equivalent of two genomes a day), a feat unthinkable a few years ago. Over its three-year lifespan, the 1000 Genomes Project will have generated 60 times the amount of sequence data that had been deposited in public DNA databases over the past quarter century. It will constitute a major bioinformatics and statistical resource for researchers and will inevitably drive innovations in data analysis and interpretation.

These examples show that large-scale, international initiatives and investment can create critical infrastructure and provide tangible and quantifiable returns on investment. Government funding and support will necessarily play a large role in the development of research infrastructures for marine biotechnology.

The success of the human genome megaprojects inspires hope that marine biotechnology megaprojects may also be successful. One of the first of these, sequencing of the tiger pufferfish, was initiated in 1989 to inform the functional genomics research for the human genome (Aparicio et al., 2002). More recently, commercially important aquaculture species have catalysed large-scale international collaboration and resulted in the mapping or sequencing of salmon (Davidson et al., 2010), trout (Palti et al., 2011) and cod (Star et al., 2011). Marine microbes have also been the focus of significant, often international, initiatives.[11]

Questions about the merit and feasibility of marine biotechnology megaprojects nonetheless remain. "What question or questions should drive such projects?" "What costs and project duration should be anticipated, with what risks?" "How might international collaboration be organised?" "Are the barriers to such megaprojects too great at present, owing to concerns about the complexity of the marine environment or the structuring of international collaboration?"

Several considerations increase the need to understand the essence of the marine biotechnology challenge and will affect the scope and focus of a potential megaproject. The challenge is first to understand better the huge, largely unstudied, complexity of marine organisms and marine ecosystems. It is close to impossible to understand the totality of marine organisms and ecosystems. At the same time, it is not possible to focus on one organism –

this is not a "human genome" project – so it will be necessary to consider the type and volume of information required for a marine biotechnology project. It will then be necessary to consider the type of research infrastructure required and how it might be developed, for example in relation to infrastructure being developed in other fields or by other megaprojects.

The development of the infrastructure brings its own questions, such as: "Who will champion the project?" or "How can the project garner the support it needs on an international level?" Many organisations have some jurisdiction in terms of monitoring and managing the ocean, some of them with significant resources and funding and some with very little. However, there is no single authoritative body able to focus the interests and resources of these fragmented, often competing, organisations and states on a megaproject.

In the absence of such an entity, states and stakeholders will need to work together to define project scope and concentrate resources. Successful megaprojects have focused first and foremost on knowledge generation at a fundamental, application-neutral level (e.g. sequencing the human genome, mapping human genetic variation). For a marine-based project, biodiversity – the core of productivity and sustainability – might be a suitable analogy.

The project would also need to compete for resources and financial support with other government initiatives and international megaprojects. Here again, the benefits of a sole administrator championing the need for research and international attention are clear. In the absence of a single champion, stakeholders will need to raise awareness of the potential opportunities of a project.

Conclusions

The renewal of interest in marine biotechnology in many OECD countries today owes much to advances in science and technology over the last decade. These advances are enabling good access to marine organisms, especially the wealth of microscopic marine microbes that seem to hold so much potential for marine biotechnology. DNA sequence information about these microorganisms can be generated faster and more cheaply than ever before, at a rate which exceeds the capacity to analyse it. In the most general sense, two main challenges are shaping the infrastructure needs of the field.

The first concerns the paucity of data about marine organisms, which makes it very difficult to characterise or classify community structure and function. Without taxonomic information or the association of genetic data and phenotype data, it is difficult to assess the potential of an organism or its genetic repertoire. For invisible microbes and larger multicellular organisms, there is a need for model systems that will allow top-down and bottom-up studies of gene expression at the individual, population, community and ecosystem levels. The resulting data will require infrastructure for coupling genetic, species and habitat information. To be most effective, global data and information infrastructures to develop and distribute data are needed. Such databases must be permanent and provide all users free and open access to all data.

The second challenge relates to the lack of analytical platforms to process data efficiently and effectively and keep pace with upstream data generation. This, along with the complexity of samples (e.g. metagenomes) to be processed, is creating the need for new e-infrastructures (e.g. novel data management platforms and information services, generation of visual analytics and web service products) to allow for better analysis across databases, especially those containing different types of information. This meta-analysis will be critical for achieving a better understanding of the biodiversity of marine ecosystems. Properly integrated into time-variable forecasts and monitoring strategies, it might also prove useful in diagnosing and predicting environmental responses to natural and anthropogenic disturbances.

Such resources will require significant investment and will benefit from multi-country and multi-stakeholder collaboration. While various international partnerships address some aspects of marine biotechnology, few international initiatives are working on these challenges. Current efforts in this regard are generally led by small groups or individual countries. There is a need to look for synergies among these projects to reduce the likelihood of duplication.

Fortunately, it is likely that scientific progress in other fields will diffuse to marine biotechnology. This has been the case for next-generation sequencing technology arising from investment in the Human Genome Project, and may well be the case for advances arising from the 1000 Genomes project. However, there are some unique features of marine biotechnology (its size, its complexity and the untapped diversity of the majority of marine organisms) that may argue for a megaproject centred on marine organisms. Details of such a project remain to be defined but could be of significant value to governments wanting to improve the productivity and sustainability of marine biotechnology.

Notes

1. For a review of the state of the art, see Liu et al. (2012), Shokralla et al. (2012) and Pareek et al. (2011).

2. www.marinespecies.org

3. www.iobis.org/

4. This problem is the focus of much attention by researchers, but the private sector has recognised the challenge (and perhaps an opportunity): in 2012 Amazon's cloud computing unit, Amazon Web Services, announced it would store the entire contents of the National Institutes of Health's 1000 Genomes Project, and provide free access to everyone. See http://bits.blogs.nytimes.com/2012/03/29/amazon-web-services-big-free-genetic-database/?scp=1&sq=Genomes&st=cse.

5. www.cebitec.uni-bielefeld.de/groups/brf/software/portal/portal/mge/?cookie_test=1.

6. Annotation of genomic sequences, or DNA annotation, is the process by which supplementary information about the sequence (e.g. genomic position to intron-exon boundaries, regulatory sequences, repeats, gene names and protein products) is added prior to inclusion in a database.

7. Information services include:

   - IMG: http://img.jgi.doe.gov/cgi-bin/m/main.cgi
   - CAMERA: http://camera.calit2.net/
   - GenGIS: http://kiwi.cs.dal.ca/GenGIS/Main_Page
   - COML: www.coml.org/
   - OBIS: www.iobis.org/
   - MetaCyc: http://metacyc.org/
   - ORION: www.orionprogram.org/OOI/
   - GEO: www.earthobservations.org/index.shtml
   - Gensc: http://gensc.org/gc_wiki/index.php/Main_Page
   - Network Workbench: http://nwb.cns.iu.edu/about.html
   - CIRCOS: http://circos.ca/
   - MIZBEE: www.cs.utah.edu/~miriah/mizbee/
   - Microbial Ecological Genomics DataBase (MegDB) www.megx.net

8. Commission of the European Communities (2008), "Communication from the Commission to the European Parliament, the Council, the European Economic and Social Committee and the Committee of the Regions Towards Joint Programming in Research: Working together to

tackle common challenges more effectively", COM(2008) 468 final, http://ec.europa.eu/research/press/2008/pdf/com_2008_468_en.pdf, accessed March 2012.

9. Full title: Joint Programming Initiative "Healthy and Productive Seas and Oceans".

10. In the first phase, 2008-10, the genomes of 1 000 individuals were sequenced to 4x coverage. Throughout all phases of the project, the genomes of about 2 500 unidentified individuals from about 25 populations will be sequenced (The 1000 Genomes Project Consortium, 2010).

11. For example, the Microbial Genome Sequencing Project is sponsored by the Gordon and Betty Moore Foundation's Marine Microbiology Initiative, http://camera.calit2.net/microgenome/, accessed September 2012.

References

Aparicio, S., J. Chapman, E. Stupka, N. Putnam, J.M. Chia, P. Dehal, A. Christoffels, S. Rash, S. Hoon et al. (2002), "Whole genome shotgun assembly and analysis of *Fugu rubripes*", *Science* 297(5585):1301-10, 23 August.

Battelle Technology Partnership Practice (2011), "The economic impact of the human genome project. How a $3.8 billion investment drove $796 billion in economic impact, created 310,000 jobs and launched the genomic revolution", Battelle, Columbus OH, May.

Benson, D.A., I. Karsch-Mizrachi, K. Clark, D.J. Lipman, J. Ostell and E.W. Sayers (2012), "GenBank", *Nucleic Acids Research* 40 (D1): D48-D53. doi: 10.1093/nar/gkr1202.

Chisholm, S.W. and C. Cary (2001), "Ecological genomics: the application of genomic sciences to understanding the structure and function of marine ecosystems", In Report, NSF workshop on Marine Microbial Genomics 1–20 (Univ. Press).

COML (Consensus on Marine Life) (2010). "First Census of Marine Life 2010: Highlights of a Decade of Discovery", COML, Washington DC.

Davidson, W.S., B.F. Koop, S.J.M. Jones, P. Iturra, R.Vidal, A. Maass, I. Jonassen, S. Lien and S.W. Omholt (2010), "Sequencing the genome of the Atlantic salmon", *Genome Biology* 11:403. http://genomebiology.com/2010/11/9/403.

Dufresne, A., M. Salanoubat, F. Partensky, F.M. Artiguenave, I. Axmann, V. Barbe, S. Duprat, M.Y. Galperin, E.V. Koonin and F. Le Gall (2003), "Genome sequence of the cyanobacterium Prochlorococcus marinus SS120, a nearly minimal oxyphototrophic genome", *Proceedings of the National Academy of Sciences*, 100(17): 10020–10025, 19 August. doi: http://dx.doi.org/10.1073/pnas.1733211100.

Kodama, Y., M. Shumway and R. Leinonen (2012), "The sequence read archive: explosive growth of sequencing data", *Nucleic Acids Research* 40 (D1): D54-D56. doi: http://dx.doi.org/10.1093/nar/gkr854.

Lerman, J.A., D.R. Hyduke, H. Latif, V.A. Portnoy, N.E. Lewis, J.D. Orth, A.C. Schrimpe-Rutledge, R.D. Smith, J.N. Adkins, K. Zengler and B.O. Palsson (2012), "*In silico* method for modelling metabolism and gene product expression at genome scale", *Nature Communications* 3;3:929, July. doi: http://dx.doi.org/10.1038/ncomms1928.

Liu L., Li Y., Li S., Hu N., He Y., Pong R., Lin D., Lu L. and M. Law (2012), "Comparison of next-generation sequencing systems", *Journal of Biomedecine and Biotechnology* 2012:251364. Epub 2012 Jul 5.

Markowitz, V.M., F. Korzeniewski, K. Palaniappan, E. Szeto, N.N. Ivanova, N.C. Kyrpides et al. (2006), "An Experimental Metagenome Data Management and Analysis System", Lawrence Berkeley National Laboratory: Lawrence Berkeley National Laboratory, http://escholarship.org/uc/item/1r9637xm/.

Markowitz, V.M., I.M.A. Chen, K. Palaniappan, K. Chu, E. Szeto, Y. Grechkin, A. Ratner, B. Jacob, J. Huang, P. Williams, M. Huntemann, I. Anderson, K. Mavromatis, N.N. Ivanova and N.C. Kyrpides (2012), "IMG: The integrated microbial genomes database and comparative analysis system", *Nucleic Acids Research* 40 (D1): D115-D122. doi: http://dx.doi.org/10.1093/nar/gkr1044.

Mushegian, A. (1999), "The minimal genome concept", *Current Opinion in Genetics and Development* 9(6):709-14.

Mora C., D.P. Tittensor, S. Adl, A.G.B. Simpson and B. Worm (2011), "How Many Species Are There on Earth and in the Ocean?", *PLoS Biology* 9(8): e1001127. doi: http://dx.doi.org/10.1371/journal.pbio.1001127.

OECD (2010), *The OECD Innovation Strategy: Getting a Head Start on Tomorrow*, OECD Publishing. doi: http://dx.doi.org/10.1787/9789264083479-en.

Palti, Y., C. Genet, M.C. Luo, A. Charlet, G. Gao, Y. Hu, C. Castano-Sanchez, K. Tabet-Canale et al. (2011). "A first generation integrated map of the rainbow trout genome", *BMC Genomics* 12:180. doi: http://dx.doi.org/10.1186/1471-2164-12-180.

Pareek, C.S., R. Smoczynski, A. Tretyn (2011), "Sequencing technologies and genome sequencing", *Journal of Applied Genetics*. 52(4):413-35, November. doi: http://dx.doi.org/10.1007/s13353-011-0057-x.

Shokralla, S., J.L. Spall, J.F. Gibson and M. Hajibabaei (2012). Next-generation sequencing technologies for environmental DNA research. *Molecular Ecology* 21(8):1794-805, April. doi: http://dx.doi.org/10.1111/j.1365-294X.2012.05538.x.

Star, B., A.J. Nederbragt, S. Jentoft, U. Grimholt, M. Malmstrom, T.F. Gregers, T.B. Rounge, J. Paulsen, M.H. Solbakken, A. Sharma et al. (2011), "The genome sequence of Atlantic cod reveals a unique immune system", *Nature* 477, 207–210, 8 September. doi: http://dx.doi.org/10.1038/nature10342.

Stratton, Campbell and Futreal (2009), "The Cancer Genome", Nature 458:719-724.

Su C., Lei L., Duan Y., Zhang K.Q. and Yang J. (2012) "Culture-independent methods for studying environmental microorganisms: methods, application, and perspective", *Applied Microbiology and Biotechnology*, Feb;93(3):993-1003. Epub 2011 Dec 22.

The 1000 Genomes Project Consortium (2010), "A map of human genome variation from population-scale sequencing", *Nature*, 467(7319):1061-1073.

Wright, J.J., K.M. Konwar and S.J. Hallam (2012), "Microbial ecology of expanding oxygen minimum zones", *Nature Reviews Microbiology* 10, 381-394, June. doi: http://dx.doi.org/10.1038/nrmicro2778.

Zhang, K., A.C. Martiny, N.B. Reppas, K.W. Barry, J. Malek, S.W. Chisholm and G.M. Church (2006), "Sequencing genomes from single cells by polymerase cloning", *Nature Biotechnology* 24, 680-686, doi: http://dx.doi.org/10.1038/nbt1214.

Chapter 5

Science, society and industry: Working together for sustainability

This chapter looks at some of the broader social challenges and opportunities relating to the conservation and sustainable development of marine bio-resources. It examines the interactions of science, industry and society that will affect this development. It focuses particularly on social issues requiring particular attention: those originating from the complexity of developing marine bioresources and from the shared and dynamic nature of the marine environment.

Science and society: Towards sustainable development

Access and sharing of benefits

Marine biotechnology raises a number of ethical, legal and social issues (ELSI). Many relate to the harvesting of marine bioresources[1] which may be widely, variably and fluidly distributed. Their harvesting, or appropriation, for biotechnological applications therefore raises questions of national sovereignty and the sharing of benefits and intellectual property rights. Many of these issues are addressed in various international instruments,[2] which seek a balance between the industrial drive for innovation and profitability and the need to ensure the sustainable use, and an equitable sharing of the benefits of, marine bioresources (Box 5.1).

It is increasingly recognised that there is a vast market for marine bio-resources (both as biomass and as a genetic resource). As a result, many companies, governments, researchers and other actors wish to have access to these resources. While there is general agreement on broad principles of prior informed consent as a basis for equitable sharing of benefits, views differ on what is equitable, the nature of benefits, and the mechanism(s) for capturing them.

What is equitable?

The 1993 Convention on Biological Diversity (CBD) is an international legal framework, which is strengthened by the Nagoya Protocol (2010)[3] (see Box 5.1), for the acquisition and use of genetic resources and associated traditional knowledge, innovations and practices of local and indigenous peoples. The CBD's Access and Benefit Sharing (ABS) negotiations resulted in a legal obligation for CBD parties to ensure fair and equitable sharing of the benefits accruing from the utilisation of genetic resources with the country that provided the terrestrial or marine genetic resources.

The principle was also extended to "local" providers or custodians of biodiversity that may have traditional knowledge associated with genetic resources or an established right to provide access to genetic resources. While further clarity is needed regarding who may constitute a "local" provider or custodian, sharing benefits with such providers is important for the equitable development of marine genetic resources, particularly as it may encourage conservation and sustainable use.

Box 5.1. The Convention on Biological Diversity's benefit sharing provisions relating to marine genetic resources

Convention on Biological Diversity (1993)

CBD-ABS negotiations were driven primarily by equity and the aim to redirect benefits back to provider(s) of genetic resources. Directing benefits back to providers was expected to create incentives to conserve biodiversity

The CBD's ABS provisions

Three fundamental access-related principles (Article 15):

- Sovereign rights over natural resources: Art.15(1)
- Prior informed consent (PIC): Art. 15(5)
- Mutually agreed terms (MATs), including the sharing of benefits arising from the commercial and other utilisation of genetic resources: Art. 15(4) and (7)

Six fundamental benefit-sharing obligations:

- Research and development results: Art. 15(7)
- Commercial or other benefits derived from use: Art. 15(7)
- Access/transfer of technology using genetic resources: Art. 16(3)
- Participation in biotechnological research on genetic resources: Art. 19(1)
- Priority access to results/benefits arising from biotechnological use: Art. 19(2)
- Traditional knowledge associated with genetic resources: Art. 8(j)

Nagoya Protocol (adopted 2010)

Addresses ABS implementation challenges not fully addressed by CBD by providing a legal framework to operationalise the CBD's third objective and Article 15.

Objective: Ensure benefits arising from utilisation of genetic resources are shared fairly and equitably

Nagoya Protocol Innovations Scope and Access Measures

Scope

- Genetic resources within scope of CBD and benefits arising from their utilisation
- Clear application to biochemical compounds: utilisation of genetic resources; derivatives

Access-related measures

- Legal certainty, clarity and transparency
- Permit or equivalent

Nagoya Protocol innovations: benefit-sharing measures

Benefits to be shared on MATs: Art. 5:

- Utilisation, subsequent applications and commercialisation

Monetary and non-monetary benefits (Annex)

- Access fees
- Milestone payments, licence fees, royalties
- Technology transfer
- Sharing research results
- Effective research participation

Box 5.1. The Convention on Biological Diversity's benefit sharing provisions relating to marine genetic resources (cont'd)

Nagoya Protocol innovations: compliance measures

Supporting compliance with provider country's domestic ABS requirements: Art. 15

Facilitating dispute resolution when non-compliance with MATs (contractual terms): Art. 18

Monitoring use: Art. 17:

- Designate "check points" to collect information at any stage of value chain (research, development, innovation, pre-commercialisation or commercialisation)
- Internationally recognised certificate of compliance
- Encouraging model contractual clauses and codes of conduct: Arts. 19 & 20

Source: CBD Secretariat (2012), www.cbd.int/.

Many parts of the world have bodies of knowledge, practices and beliefs regarding the use of biological natural resources, e.g. for curing ailments, which have evolved and been passed down through generations. Demunshi and Chugh (2010) cite examples from various countries and indigenous peoples. This knowledge, such as the use of snail flesh in curing asthma, tuberculosis, stomach disorders and eye-related problems by the tribes of Nagaland, India, could provide important starting points or development opportunities for marine biodiscovery (Box 5.2).

Marine biodiscovery is already taking place in the open ocean and the international seabed, marine areas beyond the limits of national jurisdiction, and is expected to increase. The CBD and Nagoya Protocol only apply to the continental shelves and exclusive economic zones (EEZs) of their contracting parties, and therefore do not address marine genetic resources obtained from these areas. At present, such resources are accessible to anyone for any purpose. There is no formal obligation to share benefits with the international community, although the United Nations Convention on the Law of the Sea (UNCLOS) does have provisions on international co-operation and the exchange of results with respect to marine scientific research in these areas (Glowka, 2010). While it is important to reach a shared understanding of "countries providing genetic resources" and "local" providers and a better understanding of the role and value of traditional knowledge, particularly to ensure legal certainty in the discovery and development of marine bioresources, the equitable development of marine genetic resources in areas beyond national jurisdictions may provide event greater challenges.

Box 5.2. Involvement of traditional knowledge in access and sharing of benefits

The Suriname project

This project is concerned with the medicinal qualities of the coastal forest plants of Suriname. The US government relies on access and sharing of benefits between companies and the Maroon tribes of the region of Samaraka in South America (Guerin-McManus et al., 1998).

Unilever's Best Food

This project includes an agreement (a kind of bioprospecting contract) between Unilever's Best Food and the local people of Vietnam. According to the agreement nuoc mam (fish sauce) is produced from anchovies by the Phu-Quoc islanders in a traditional manner and the finished sauce is bottled, packed and marketed by Unilever. The islanders are free to sell their product in the wholesale market at Unilever's equivalent market price (Kazmin, 2003).

Bioprospecting in the South Pacific

The University of South Pacific, along with what is now Glaxo SmithKline and the Fijian Affairs and Fisheries Department, developed a draft ''biodiversity access and benefit sharing policy'' which resulted in the establishment of a successful bioprospecting enterprise. The enterprise has provided licensing fees of USD 30 000 to a trust fund established by the community.

National Biodiversity Institute (INBio)

Through an agreement between the National Biodiversity Institute (INBio), a non-profit scientific organisation in Costa Rica, and Merck, INBio will provide 10 000 samples of plants, animals and soil to Merck, which will have the exclusive right to study these samples for two years. Merck will retain the patents to any drugs developed using the samples but will pay royalty fees for those patents to INBio. Merck will also pay INBio USD USD 1 million up front and will give the institute an additional USD 130 000 worth of laboratory equipment (Coughlin, 1993).

Source: Demunshi and Chugh (2010).

Marine gene patents (even if they do not result in exploitation of bioresources) provide an indication of access to marine bioresources and illustrate the challenge. A search of the patent division of GenBank from 1999-2009 identified 677 international claims of marine gene patents. These patents originated from only 31 of the world's 194 countries. Some 90% of these patents belonged to just ten countries[4] which account for only 20% of the world coastline (Arnaud-Haond et al., 2011). These ten countries have the scientific and technological capacity to explore and develop resources, which many other countries, especially developing coastal countries, lack. If this issue is not addressed, successful biodiscovery, of which patents are one indicator, may spur further investment by capacity-rich countries and create further inequities.

In discussions under the auspices of the United Nations General Assembly,[5] some countries argue that marine genetic resources in areas beyond the limits of national jurisdiction should be viewed as a common heritage of humankind. Other countries take a different view, considering the international legal situation as settled. Ongoing discussions would benefit from identification of the actual or potential scientific/economic/ commercial value of marine genetic resources in areas beyond national jurisdictions, the profiling of the sectors involved (including their practices of exchange and use), an assessment of the accessibility of marine genetic resources (*in situ*, *ex situ* and *in silica*) and related data, and an assessment of the ability of developing countries to access and utilise marine genetic resources and related data (Glowka, 2010).

As work proceeds to identify situations in which sharing of benefits is required or might be desirable, questions arise concerning the basis for determining the sharing of particular benefits in particular situations. While the Convention on Biological Diversity emphasises sharing of benefits, none of the international and national legal instruments determines the extent of the sharing of profits or defines benefits for the indigenous communities that provide traditional knowledge to companies interested in product commercialisation. It seems that one of the first questions regarding access and sharing of benefits is how to value the contributions of different stakeholders in the discovery and development process.

Understanding benefits

The CBD and the Nagoya Protocol embody a broad international consensus that the sharing of benefits is a necessary part of the equitable development of bioresources. However, as such benefits can be defined in different ways, it will be important to specify what kinds of benefits marine biotechnology can offer. The Nagoya Protocol and the FAO International Treaty on Plant Genetic Resources for Food and Agriculture (ITPGRFA)[6] may provide a basis for addressing this issue. The Nagoya Protocol reproduces, in an annex, an indicative list of monetary and non-monetary benefits derived from the CBD (Secretariat of the Convention on Biological Diversity Guidelines, 2002), and the FAO International Treaty (which was negotiated to be consistent with the CBD) identifies four main forms of benefits: *i*) the exchange of information; *ii*) access to and transfer of technology; *iii*) capacity building; and *iv*) financial and commercial flows.

Financial benefits, which relate to commercialisation (e.g. royalties or cash exchanges), are often the first type considered. While equitable sharing of financial benefits is important, it can be short-sighted to focus solely on these, since commercial applications of marine bioresources are typically not immediately apparent and achieving a commercialisable product may take many years. Therefore, financial considerations should not overshadow aspects such as the sharing of genomic or taxonomic information, capacity-building or technology transfer. In some cases, access to and transfer of technology or capacity building may confer greater net benefits in the short to medium term by enabling provider countries and local communities to take advantage of marine bioresources and may allow them to negotiate fair and balanced deals over the longer term.

The focus on access and sharing of benefits should not overshadow the importance of conservation and sustainable use of marine bioresources, as this may determine the possibility of deriving future benefits. The CBD clearly recognised the importance of this. It requires its contracting parties to facilitate access to genetic resources for environmentally sound uses [Art. 15(2)]. For example, Costa Rica has realised monetary and non-monetary benefits from access to its genetic resources while supporting programmes aimed at conserving its biodiversity. It benefits financially through a direct contribution to research budgets and royalties, in addition to transfers to public universities, research infrastructure and research funding. Of these benefits, 10% of research budgets and 50% of royalties are channelled to conservation work. Non-monetary benefits include the development of scientific networks for R&D programmes, sharing of results, technology transfer, capacity building, publication and dissemination of data relevant to conservation and sustainable use of biodiversity and establishment of *ex-situ* collections.

Mechanisms for sharing benefits

Two broad approaches to sharing benefits have emerged: case-by-case and multilateral agreements. A case-by-case approach involves negotiations and agreement between parties concerning access to genetic resources and the fair and equitable sharing of benefits derived from their utilisation. As countries implement the CBD, many have taken this approach. They are likely to continue to do so as they implement the Nagoya Protocol, although neither instrument precludes multilateral approaches.

A multilateral approach is attractive when negotiating benefit-sharing arrangements that may involve high transaction costs, e.g. when it is difficult to identify the source(s) of genetic resources or when a resulting product implies contributions from many sources over time. This may be the case for genetic resources for food and agriculture and may have important implications for food security. The FAO ITPGRFA has used a multilateral approach to facilitate access to an internationally agreed group of plant crops important for food security (e.g. wheat) which are in the public domain and managed by governments. The ITPGRFA creates a common pool of plant genetic resources located *ex-situ* for the purposes of research, conservation and breeding. Access to the pool is governed by a standard material transfer agreement (sMTA). Both the treaty and the sMTA provide for the possibility of monetary and non-monetary sharing of benefits, with facilitated access to the common pool considered an important benefit. A multilateral approach to access and benefit sharing may be useful for non-sedentary marine genetic resources which are not within the exclusive domain of any one country. For these resources, a "commons" approach that recognises collective interests of access and sharing may be more useful.

Multilateral approaches may also be useful when there is a disparity between activities in exclusive economic zones (EEZ) and the geographic distribution of resources in the region. This situation may arise for regional seas. For the Mediterranean basin, for example, targeting marine genetic resources as a common resource, perhaps via the development of the Mediterranean Science Commission (CIESM) Marine Peace Parks, has been suggested as a possibility. Beyond the limits of national jurisdictions, marine genetic resources might be considered a common pool; this would call for consideration of multilateral approaches to access and benefit sharing. Such approaches will need to take into account a multiplicity of stakeholders with different interests and levels of technological and economic development. They will likely require new types of legal instruments that capture direct and indirect benefits at societal level, while creating incentives for research and development.

When developing mechanisms for sharing benefits, information on the spatial distribution of marine genetic resources and on the geographic location of sampling activities will be needed to help identify the applicable legal regime. The maritime zones delineated by UNCLOS will need to be kept in mind. This will help to identify who should benefit. For example, there is the question of whether a geographic region can be considered a stakeholder for purposes of benefit sharing. If so, it will be important to understand how the marine genetic resources can be shared, how the countries of the region interact and whether appropriate governance mechanisms are in place to capture benefits at regional or national levels. It will also be important to identify the major stakeholders in each of the relevant states.

Finally, sharing of the benefits of marine genetic resources from areas beyond national jurisdictions raises questions about the governance of such areas and about the overall authority for access and sharing of benefits. The treatment of marine genetic resources may be influenced by geopolitical or economic conditions in individual countries and there may be a need for some common benchmarks for governance of these resources. UNCLOS can provide a framework, but its provisions may be insufficient to resolve all relevant issues and it may require further elaboration.

Social engagement

Many of the ocean's beneficial functions and services stem from the interconnected ecosystems of marine bioresources distributed across an immense shared environment that can be positively or negatively affected by the actions of countries and stakeholders. It is increasingly evident that stakeholders need a common understanding of the ocean and of the economic, social and environmental aspects of the sustainable development of its resources. This will involve dialogue and diverse forms of engagement by all stakeholders, including developers and users of innovations and those relying on the ocean for other purposes. There is already considerable evidence of the convergence of the views of the international community on several aspects of protection of the marine environment. A first indication was the entry into force of UNCLOS in 1994. Its general provisions provide a foundation for marine environmental protection. These have been complemented and deepened by those of the CBD.[7] The Convention recognises the conservation of biodiversity as a "common concern of humankind".

The CBD, its programme of work on marine and coastal biodiversity and, more recently, the adoption of the Biodiversity Strategic Plan (2011-20), which includes the 20 Aichi Biodiversity Targets, provide a strong framework for the conservation and sustainable use of marine biodiversity. The CBD's work on the biodiversity of the deep seas and open oceans, particularly the adoption by its Conference of the Parties in 2008 of scientific criteria for identifying ecologically or biologically significant areas in the global marine realm, has spawned the Global Ocean Biodiversity Initiative (GOBI). Established in 2008, GOBI is an international partnership for advancing the scientific basis for the conservation of biological diversity in the deep seas and open oceans. It aims to support countries, as well as regional and global organisations, use and develop data, tools and methodologies to identify ecologically or biologically significant areas, building on the CBD's scientific criteria for ecologically or biologically significant areas. The initial focus of GOBI is on areas beyond national jurisdiction.

Marine protected areas can conserve and make sustainable use of marine biodiversity. They can help protect the contributions of marine biodiversity to human well-being either directly or in the form of ecosystem goods and services (e.g. cultural, recreational). The management of such areas typically places restrictions on human activities in certain regions to safeguard the natural environment and conserve biodiversity. In 2010, when marine protected areas were first included in the World Database on Protected Areas (WDPA) they numbered more than 5 880 in 1.17% of the ocean. By mid-2012, their number reached 7 354, as nations and regions designated more areas for protection.[8] The Aichi Target[9] is conservation of 10% of coastal and marine areas, especially those of particular importance for biodiversity and ecosystem services, by 2020.

Marine protected areas can be designated and managed at different scales. For example, local fishermen in Cabo Pulmo, Mexico, faced with declining catches and the possibility of losing their livelihood, lobbied to have the area designated a national park. In 1999, four years after the establishment of the nature reserve, and with little change in biomass, the community declared and enforced no-catch zones in the park. Over the next decade, fish biomass increased by over 460%, whereas there was little or no change in other protected or open access areas (Aburto-Oropeza et al., 2011). This increase in biomass is the largest measured increase in a marine reserve worldwide and is attributed to a combination of social (strong community leadership, social cohesion, effective enforcement) and ecological factors. The recovery of fish biomass has brought significant economic benefits to the community and indicates that a bottom-up approach may be a viable response to unsustainable coastal development and fisheries collapse.

Because the ocean provides a range of critical functions and services (e.g. food, nutrient cycling and oxygen generation), diverse strategies will be needed to sustain marine biodiversity. Clearly, well-managed marine protected areas can restore or conserve endangered or threatened species. However, in many places, restrictions are poorly enforced, and work is needed to improve the effectiveness of this approach.

The appropriateness of this approach for marine microbes, or for species of which little is known, is uncertain. Protecting the habitat of marine microbes would undoubtedly require quite large reservoirs, although it is difficult to suggest the most appropriate size given the limited understanding of marine microbes and associated ecosystems. Similarly, as the roles that species play in this global system are not clear, and because of the immense difficulties of a species-focused strategy, strategies for biodiversity preservation might best focus on protection of habitat.

Stakeholders have adopted bottom-up and top-down approaches with considerable success. However, because the vast majority of the ocean, especially the open ocean, is beyond the jurisdiction of any one state, it is not clear how marine protected areas could be established or who would enforce them. This is presently being discussed by a working group of the United Nations General Assembly, which is also discussing benefit sharing as it relates to the utilisation of marine genetic resources from areas beyond national jurisdictions.

Priority setting and special social issues

The shared nature of marine bioresources will create particular challenges for setting priorities for their use and development. Consideration should be given to how priority setting can encourage the use of marine biotechnology for the national and global good. The alignment of research policy with national bioresource policy can help achieve an appropriate balance between development and sustainability. A number of governance issues from the laboratory to commercialisation (e.g. biosecurity, biosafety, public perception) are also likely to require policy attention.

Public engagement will be necessary to stimulate development of the field and to advance the policy agenda. It will be important to have an ongoing, inclusive dialogue on the opportunities offered by marine biotechnology and their environmental implications, and for this dialogue to take place at regional, national and international levels. The goals of economic productivity and wealth creation need to be seen in terms of the cultural and social well-being not only of coastal populations but of the entire world.

Specific applications of marine biotechnology may be associated with social issues that should be addressed in a particular context. For example, *Aquaculture, Innovation, Social Transformation* (Culver and Castle, 2008) looked at issues associated with modern aquaculture, such as animal welfare, intellectual property, environmental sustainability and the interactions between traditional and local knowledge and modern aquaculture as they relate to the social effects of intensified fish farming and production, food security, environmental sustainability and global competitiveness.

Issues such as these need to be considered in relation to other areas of marine biotechnology and its application and to other uses of the marine environment (e.g. oil and gas extraction, transport, recreation, aquaculture, culture, etc.) which affect the extent to which marine biotechnology contributes to the bioeconomy and economic growth.

Climate change and many of the ensuing ecosystem changes, such as habitat loss, the rise of invasive species and population decline, are all likely to raise the profile of many of the ethical issues surrounding the sustainable development and exploitation of marine resources.

Collaboration with industry for knowledge co-creation and translation

The last two decades have seen an opening up of the innovation process. It is increasingly recognised that many actors participate in innovation processes and organisations increasingly rely on external sources of knowledge or innovation via inter-organisational network relationships. As the OECD observed in 2008:

> Globalisation has changed the location and way in which knowledge is generated. Confronted with increasing global competition and rising research and development (R&D) costs, companies can no longer survive on their own R&D efforts but look for new, more open, modes of innovation. Companies' innovation activities are increasingly international, and they are embracing "open innovation" – collaborating with external partners, whether suppliers, customers or universities, to keep ahead of the game and get new products or services to market before their competitors. At the same time, innovation is being "democratised" as users of products and services, both firms and individual consumers, increasingly become involved in innovation themselves. (OECD, 2008)

As in other fields, effective partnerships will play a large role in developing marine biotechnology and in translating new scientific and technological knowledge into social and economic benefits. Successful innovation will require partnership with stakeholders throughout the innovation cycle: suppliers, consumers, competitors, private R&D firms, universities and higher education institutions, and government and public research institutes. It will be important to understand the types of partnerships that facilitate innovation and the mechanisms that are effective for initiating and supporting them.

Many organisations have recognised the benefits of viewing the relation between research and industry as a partnership in which contacts between researchers and industry accelerate the creation and application of knowledge. Industry-university partnerships now take a range of forms: collaborative research, university-industry research centres, contract research and academic consulting (Perkmann, 2007). Governments have also recognised the benefits of linking organisations and external researchers earlier in the innovation process and provide a range of financial and other incentives. In Canada, federal funding agencies for science and engineering and social science have specific funding programmes for university-industry partnerships.[10] These programmes require 30-50% matching support from industry or other research partners. In Canada and elsewhere, knowledge exchange meetings foster constructive dialogue with industry and earlier engagement.[11] Further work to specify incentives for involvement and to ensure all partners benefit from the collaboration would be valuable.

In marine biotechnology also, interaction between researchers and industry can provide opportunities for co-development or sharing of databases and other infrastructure to support basic research. Currently, a number of databases or culture collections developed with public funding are shared with industry. Established in 1995 and housed in the Department of Aquaculture, Pukyong National University, Korea, the Marine Microalgae Culture Centre maintains cultures from microalgal species collected from Korean coastal waters which are shared with researchers in universities, research institutes and industry. The European Marine Biological Resource Centre (EMBRC), discussed in Chapter 4, provides small- and medium-sized enterprises (SMEs), academia and industry with access to marine biodiversity, associated metadata and extractable products, for their marine biotechnology projects. Platforms such as databases and biobanks provide an excellent focus for collaboration and the open sharing of data and data products. Yet, while industry has access to databases developed with public, and sometimes private, funding, the converse may not always be true. Consideration should be given to incentives for industry and scientists to work together in this area.

Partnerships between researchers and industry have had positive results (e.g. sharing of databases) but have also encountered challenges. One is the timing of the engagement between researchers and industry. Engagement with industry is often regarded as incidental to basic R&D or as a post-research, downstream activity. This can leave R&D results stranded, either without a ready market or unable to reach the anticipated market for technical or feasibility reasons. Earlier collaboration with industry can help to ensure that the products of marine biotechnology research are suitable for scaling up to industrial production.

Lessons regarding the timing and possible mechanisms for effective linkages between researchers and industry can be found in programmes of the European Union and activities of OECD member countries. For example, the Algae Technologies (BIOFAT)[12] project, funded in large part by the European Commission's Seventh Framework Programme BIOfuel, involves nine academic, industrial and public-sector partners and aims to demonstrate the economic viability and environmental sustainability of biofuels derived from macroalgae on a large scale.

In France, GREENSTARS is a project that links academic research on microalgae with industrial partners. It targets markets in biofuels and animal food, cosmetics, green chemistry and energy. The initiative is led by the Institut National de la Recherche Agronomique (INRA) in collaboration with 45 partners from the public sector, SMEs, multinationals, local authorities, and competitiveness clusters. Its aim is an integrated biorefinery for biofuels and high value added substances using microalgae fed with industrial emissions and organic wastes. The initiative has a budget of EUR 160 million for 10 years, of which roughly 20% from public grants. By 2016, the partnership will have industrial prototypes based on state-of-the-art technologies that will enable the building of a viable economic and environmental model.

In industrial biotechnology, including the expanding area of renewable or bio-based fuels, industry is generally involved at an earlier stage of the R&D process than in other sectors. As a result, more commercially significant inventions originate from non-academic research (Mowery and Sampat, 2005). In the biotechnology and pharmaceutical sectors, instead, advances in university research affect industrial innovation more significantly and more directly than in other sectors. In these sectors, industry typically is involved later in the innovation process and a significant portion of the innovation process, and of the related the R&D expenses, is shifted to taxpayers (via universities and public research institutes). Different fields thus have different patterns of engagement. Mechanisms and incentives may need to be adjusted for the emerging organisations that use marine biotechnology in commercial and non-commercial applications.

While marine biotechnology may well follow an innovation path similar to that of other biotechnology sectors, this is not certain. It will be important to plan for the possibility that it will require different partnership approaches and support. As it is a relatively new area, it may be particularly important to support a large amount of basic research (relative to applied research) on marine microbes and the functioning of complex marine ecosystems.

Incentives or other support may be required to encourage academics and other actors in basic research to participate in the full innovation cycle up to commercialisation. To achieve an appropriate balance between basic and applied research in advancing marine biotechnology will also require business models for developing and producing marine biotechnology products and services that ensure the right incentives and support.

Stakeholder engagement for diffusion of innovation

The preceding discussion has shown the importance of engagement and dialogue with a broad range of stakeholders, including local custodians and end-users of marine resources. The role of early links between researchers and private-sector actors in enabling innovation and diffusion in the marketplace has also been explored. Similarly, engagement with other stakeholders is important for removing barriers that may affect how innovations reach end users. Earlier discussions with regulators, for example, may help to reduce the risks of investment in R&D by ensuring that appropriate governance frameworks are in place so that an innovation can reach the market. Such discussion can also contribute to the development of appropriate biosafety, waste disposal or other standards that affect the diffusion of a given innovation. For new products, life-cycle analysis may affect incentives to adopt innovations.

As the marine biotechnology field develops, it will be important to assess the types of stakeholder engagement that facilitate the development and diffusion of innovations. It will be necessary to understand the mechanisms that encourage engagement and to create the appropriate incentives for the various stakeholders.

Conclusions

The marine ecosystem provides a range of services, ranging from modulating climate change to the accumulation of carbon to nutrient recycling. Use or collection of bioresources from the marine environment will affect the marine ecosystem and may in turn affect nations' capacity to derive wealth or to address global challenges. As the field of marine biotechnology develops, it will be important to identify a governance and regulatory environment that fosters the creation of national wealth and global benefits in harmony with the protection of marine biodiversity and ecosystem health.

The Convention on Biological Diversity is evidence of the near universal consensus in the international community that marine biodiversity must be conserved and used sustainably and that the benefits derived from the use of marine genetic resources should be fairly and equitably shared. The use of marine genetic resources will be the basis of new partnerships and approaches to innovation while creating incentives to conserve and use marine biodiversity sustainably.

These efforts, while still in their infancy, show the need for broader and deeper discussions about effective approaches and governance, especially for areas beyond national jurisdictions. While these issues are being addressed,[13] there is no co-ordinated approach to the conservation, sustainable use and sharing of the benefits of marine biodiversity.

Effective earlier links between academic researchers and industry can mobilise knowledge and foster innovation based on marine genetic resources and the development of marine biotechnology. Such engagement may take many forms, from dialogue to large-scale working partnerships, and it is certain that all stakeholders – developers of innovations, those using innovations, those relying on the ocean for other purposes – should be involved. It is important to identify the most effective partnerships and the best ways to support them to enable innovations in and applications of marine biotechnology.

Finally, owing to the breadth of potential applications of marine biotechnology, government actions will affect investment in basic R&D, the development of partnerships, and the diffusion of innovations to end users. Governments will need to consider how policy settings can best enable progress and innovation in this field.

Notes

1. Marine bioresources are understood as the resources derived or originating from biological material from the marine environment. They include both biomass such as algae and fish, and marine genetic resources from these and other biological specimens such as marine microbes. This chapter focuses primarily on marine genetic resources as they relate to marine biodiversity and biodiscovery, but many of the themes explored may also be relevant to marine biomass.

2. Such as the United Nations Convention on the Law of the Sea (UNCLOS) and the Convention on Biological Diversity (CBD) and its Cartagena Protocol on Biosafety and its Nagoya Protocol.

3. The Nagoya Protocol on Access to Genetic Resources and the Fair and Equitable Sharing of Benefits Arising from their Utilisation (ABS) to the Convention on Biological Diversity is a supplementary agreement to the UN Convention on Biological Diversity. It provides a legal framework for the effective implementation of one of the three objectives of the CBD: the fair and equitable sharing of benefits arising from the utilisation of genetic resources.

4. The United States, Japan and Germany account for 70% of the patents.

5. The UNGA Informal Working Group on Biodiversity beyond the Limits of National Jurisdiction was mandated in 2001 to initiate discussion of the legal framework for the conservation and sustainable use of biodiversity in areas beyond the limits of national jurisdiction.

6. See also www.planttreaty.org, accessed August 2012.

7. Marine biodiversity was the theme of the 2012 International Day for Biological Diversity (IDB). Designation of the theme of marine ecosystems provided parties to the CBD and everyone interested in marine life the opportunity to raise awareness of the issue and increase practical action.

8. See www.protectedplanet.net/, accessed August 2012.

9. The Aichi Biodiversity Targets are part of the Convention on Biological Diversity Strategic Plan 2011-20, www.cbd.int/sp/targets/.

10. Canada's National Science and Engineering Research Council (NSERC) and the Social Science and Humanities Research Council (SSHRC) provide funding through Collaborative Research and Development Grants and Partnership Grants (respectively).

11. For example, "Where Industry meets Science", Concarneau, France, 28-29 August 2012, http://concarneau.mnhn.fr/sites/concarneau.mnhn.fr/files/upload/programme_rendez_vous_concarneau_2012.pdf, accessed August 2012.

12. www.biofatproject.eu/, accessed August 2012.

13. The UNGA Informal Working Group on Biodiversity Beyond the Limits of National Jurisdiction.

References

Aburto-Oropeza O., B. Erisman, G.R. Galland, I. Mascareñas-Osorio, E. Sala and E. Ezcurra (2011), "Large Recovery of Fish Biomass in a No-Take Marine Reserve", *PLoS ONE* 6(8): e23601. doi: http://dx.doi.org/10.1371/journal.pone.0023601.

Arnaud-Haond, S., J.M. Arrieta and C.M. Duarte (2011), "Marine Biodiversity and Gene Patents", *Science*, 331(6024):1521-1522.

Coughlin, M.D., Jr. (1993), "Using the Merck-INBio agreement to clarify the Convention on Biological Diversity", *Columbia Journal of Transnational Law* 31(2):337–375.

Culver, K. and D.G.A. Castle (eds.) (2008), "Aquaculture Innovation and Social Transformation", International Library of Environmental, Agricultural and Food Ethics.

Demunshi, Y. and A. Chugh (2010), "Role of traditional knowledge in marine bioprospecting", *Biodiversity Conservation* 19:3015–3033.

Glowka, L. (2010), "Evolving Perspectives on the International Seabed Area's Genetic Resources: Fifteen Years after the 'Deepest of Ironies' in *Law, Technology and Science for Oceans in Globalisation*, Martinus Nijhoff Publishers / Brill, Leiden/ Boston MA, pp. 397-419.

Guerin-McManus, L.M. Famolare, I.A. Bowles, S.A.J. Malone, R.A. Mittermeier and B. Rosenfeld (1998), "Bioprospecting in Practice: A Case Study of the Suriname ICBG Project and Benefits Sharing under the Convention on Biological Diversity", http://belfercenter.ksg.harvard.edu/files/pdf2.pdf, accessed March 2012.

Kazmin, A. (2003), "Unilever has a taste of success in Vietnam", www.vietnamembassy-usa.org/news/story.php?d=20031202110206, accessed 14 November 2008.

Mowery, D.C. and B.N. Sampat (2005), "The Bayh-Dole Act of 1980 and University–Industry Technology Transfer: A Model for Other OECD Governments?" *Journal of Technology Transfer*, 30 ½:115–127.

OECD (2008), *Open Innovation in Global Networks*, OECD Publishing doi: http://dx.doi.org/10.1787/9789264047693-en.

Perkmann, M. (2007). University-industry relationships and open innovation: towards a research agenda. *International Journal of Management Reviews*, 9(4):259–280.

Secretariat of the Convention on Biological Diversity (2002). Bonn Guidelines on Access to Genetic Resources and Fair and Equitable Sharing of the Benefits Arising out of their Utilisation. Montreal: Secretariat of the Convention on Biological Diversity, www.cbd.int/doc/publications/cbd-bonn-gdls-en.pdf.

Chapter 6

Conclusions: Delivering on the promise of marine biotechnology

This chapter presents conclusions, identifies the policy areas of greatest potential impact for enabling the sustainable development of marine biotechnology and highlights the challenges of supporting both productivity and ocean sustainability. It provides a foundation for future work to overcome the challenges and realise the potential of marine biotechnology.

Marine biotechnology in the 21st century

Advances in genome science and related technologies have led many countries to focus attention on marine biotechnology. Marine organisms, once difficult to access and study, can now be examined more quickly and in greater detail than ever before. Genomic and metagenomic analyses of marine organisms and the marine environment are revealing the complexity and biodiversity of marine ecosystems. By all estimates, the ocean's biodiversity far exceeds that of terrestrial environments but is largely unknown. Yet marine bioresources may be a significant source of new biological and chemical processes and principles from which new bioactive compounds can be isolated, modelled or created. Marine biotechnology is increasingly recognised as a potential source of innovation.

Addressing global challenges

Advances in genomics and computer science have transformed earlier views of the ocean. It is no longer simply a source of food and a way to transport goods but a vast reservoir of genetic potential and a means of achieving a wide range of socioeconomic benefits. The application of marine biotechnology in a number of sectors suggests that it may help to meet the global challenges of population health, food and fuel security and greener industrial processes:

- *Human health and well-being*: The ocean is recognised as a source of drugs and natural products with various functionalities. As of 2012, seven marine-derived drugs had received FDA approval, eleven were in clinical trials and 1 458 were in the pre-clinical pipeline. Marine microbes are of particular interest as new sources of antibiotics for treating drug-resistant bacterial infections.
- *Industrial biotechnology*: Novel catalysts from marine organisms are used in a number of industrial applications and the potential of algae as a sustainable biofuel and a source of refined high-value chemicals is being explored.
- *Food supply*: More sustainable food production is being achieved through new fish breeding and rearing technologies, the development of novel feeds, the understanding of health issues and new opportunities for the use of wastes.
- *Environment sustainability*: Tools and processes are being developed to monitor and address potentially negative environmental impacts.

A new source of economic growth

The economic potential of marine biotechnology appears to be significant and growing: in 2010 the market was estimated at EUR 2.8 billion with annual growth of 4-12%, depending on the model used. Growth in marine biotechnology is expected to create new jobs along the value chain from academia to industry in the marine sector and to sectors such as pharmaceuticals, food, industrial processing and nutraceuticals.

This is spurring renewed interest in marine bioresources and marine biotechnology as a source of innovation and economic growth. Many governments and regions are investing in marine biotechnology and developing frameworks to support the field. Some countries have incorporated marine biotechnology in their national "bioeconomy" strategies and blueprints.[1]

Challenges to development

This report has considered the potential of this burgeoning field as well as issues that may impede its development. In order to ensure its relevance to policy for marine biotechnology, it has focused on issues that are unique to this field. Because marine biotechnology is inextricably associated with the marine environment, particular attention must be paid to the vast physical and geographic distribution of bioresources, the diversity of these bioresources and the complexity of the marine ecosystems.

Governance of distributed bioresources

The distributed, and often dynamic, nature of bioresources means that they are often found beyond the jurisdiction of any one nation. For most purposes, they can therefore be considered shared resources. This is particularly the case for organisms in the deep ocean or open seas outside exclusive economic zones (EEZ) and for motile or current-based organisms with a wide geographic range. The Convention on Biological Diversity and the Nagoya Protocol provide a broad framework governing access to and sharing the benefits of resources within EEZ. However, this framework is less relevant and effective for organisms beyond these areas. This raises questions about ownership, access to these bioresources and the sharing of the benefits to be obtained.

Similar questions are raised concerning the protection of these shared resources. This report has discussed the utility of marine protected areas and various approaches to their governance and protection. The complexity of marine ecosystems and the comparative lack of knowledge regarding the functioning of and interaction of organisms and between them and their environment complicate efforts to protect or develop marine resources. How does the ecosystem respond to varying levels of an organism and how can a single organism or functionality be singled out? This raises two development-related challenges: the infrastructure needed to develop marine bioresources and the sustainable use of those resources and their ecosystem services.

Developing infrastructure

To protect the marine environment and to derive social and economic benefits from marine biotechnology will require in-depth knowledge of marine bioresources and their ecosystems. It will also require an appropriate R&D infrastructure to enable the generation, analysis, sharing and dissemination of knowledge about marine bioresources. Marine biotechnology has benefited significantly from prior life-science investment in R&D, particularly for human genomics. Genome sequencing is no longer the barrier it was a decade ago and our understanding of marine bioresources has improved significantly. Metagenomic analysis is providing a means to access and study the complexity of the marine environment but it also reveals deficiencies in marine biotechnology R&D, which are compounded by our limited understanding of ocean bioresources. Access to these, especially in the deep ocean, remains a challenge, and the complexity and novelty of marine bioresources makes annotation of data difficult. New infrastructures are needed, with new models, new culture systems and new bioinformatics-based tools to visualise and analyse genomics and other types of data.

Economic monitoring tools

As governments seek to incorporate marine biotechnology into innovation strategies, it is becoming clear that they lack the tools needed to measure the success of these strategies. This is due to the lack of appropriate indicators, the difficulty of tracking outputs of marine biotechnology across a wide range of sectors, and the difficulty of separating the contribution of marine biotechnology to the innovation process from the resultant products. The field differs from biomedical or industrial biotechnology in that marine biotechnology is still defined in terms of its source (or target) material, rather than the market it serves.

As governments and other stakeholders continue to invest in marine biotechnology, there will be a need for new measures and indicators to track investment and outcomes. These will need to go beyond existing measures of investment to find ways to measure outputs in different sectors. It will also be necessary to measure the non-market contribution of the ocean and marine biotechnology as the promise of marine biotechnology, both economic and otherwise, owes much to the health of the marine environment.

Strategic partnerships to enable innovation

The wide range of applications of marine biotechnology creates special challenges. There is a need, as in other fields, to form strategic partnerships for knowledge co-creation, transformation, mobilisation and diffusion from basic R&D through to diffusion of innovations in the market place. There is also a need for dialogue among all stakeholders regarding access, development and exploitation of bioresources from the marine environment. However, given the variety of sectors concerned, it is unlikely that a one-size-fits-all approach to marine biotechnology partnerships will be effective.

It is also clear that policies to support marine biotechnology applications in one sector might be quite different in others. These differences relate to differences in markets and in the knowledge or capacities of stakeholders in different sectors. It will be important to consider many different approaches to knowledge creation and sharing and the commercialisation of innovations in different sectors.

Addressing hurdles to marine biotechnology development

Throughout previous chapters, two recurrent themes have appeared important for the sustainable development of marine biotechnology: the need for communication among stakeholders and the need for internationally co-ordinated action. While these themes are not unique to the field, their emergence from this study's focus on the distinctive features of marine biotechnology makes them especially noteworthy and in need of attention in future policy work.

International collaboration and co-ordinated action

The sheer size of the ocean, which is far larger than the terrestrial surface of the planet, has long served as a call for internationally co-ordinated development of resources. This report has shown that the ocean is home to vast bioresources which are widely distributed and rarely confined to any nation's exclusive economic zone. Most are motile, self-propelled or moved by ocean currents, and present across vast geographic regions; others lie in more restricted habitats of the deep ocean and other areas beyond national jurisdiction. No single nation owns these highly distributed resources and their conservation will require internationally co-ordinated agreements and actions.

International co-operation has resulted in a framework for the conservation of biodiversity, but further co-operation is needed. Chapter 5 detailed the need to develop a framework for access to marine genetic resources in areas beyond national jurisdiction. It also highlighted the potential value of a framework for enforcing actions related to the sharing of benefits from these resources. These challenges are heightened in the absence of a single governing authority, especially as the focus turns from protecting marine resources to their exploitation and as different stakeholders engage in different ocean-related activities.

Development of the R&D infrastructure for improving understanding of marine bioresources and supporting their sustainable development can also profit from international co-operation. International megaprojects, such as the human genome project described in Chapter 4, have successfully focused considerable resources on the development of specific infrastructure. For marine biotechnology, however, next-generation sequencing has revealed gaps in the infrastructure for analysing genomic information and underscored how little is known about the complexity of life in the ocean. For marine biotechnology, the challenge will be to create infrastructure to help understand the complexity of marine organisms and related marine ecosystems. Given the type and volume of information required to enable the field and the type of research infrastructure required to obtain this information, development of the infrastructure may best be accomplished though international collaboration.

As discussed in Chapter 3, there are many opportunities for international co-ordination on the development of indicators and measures for marine biotechnology. The development of standards or measures of ocean health may best be accomplished through international collaboration as they will be most useful if universally accepted.

The development of economic statistics and indicators may need to be modelled on established national metrics. However, it would be useful to have international agreement on the definition of marine biotechnology, especially as it relates to biodiversity and to other forms of biotechnology. This may affect the development of relevant economic indicators and measures, especially those for measuring the health of the ocean.

Improved dialogue or communication

Focused and effective international dialogue will be needed to address hurdles such as development of indicators, R&D infrastructure and sustainable development of marine resources. Dialogue at national and regional level among end users, regulators, the private sector and researchers will also be important for innovation in marine biotechnology and its applications (see Chapter 5).

Consultation with civil society regarding the development of marine bioresources can help to ensure sustainable development and optimal diffusion of the technology. Early dialogue can help to secure public support for innovation in the marketplace (social acceptance) and, in cases of local custodians of marine-based genetic resources, it can expedite access to marine bioresources and downstream knowledge co-creation. The Nagoya Protocol provides a framework for discussions of access to marine genetic resources and traditional knowledge for the development of resources, but there is scope for earlier dialogue and stakeholder engagement.

The challenges associated with the scaling up of laboratory-based proof of concept to industrial level are great but may be mitigated in part through earlier collaboration between industry and academic or other researchers, including through strategic partnerships. This opportunity has been recognised and many government and other initiatives encourage industry-academic partnerships early in the process to support co-development of knowledge and platforms. However, different types of marine biotechnology applications and different industries will require different types of dialogue and will benefit from different strategic partnerships.

Finally, there may be a need, as evidenced by the emergence of marine biotechnology in national bioeconomy strategies, for a communication strategy around marine biotechnology, perhaps with a focus on environmental issues and sustainability more broadly. A communication strategy might explain what biotechnology is and how applications of marine biotechnology can affect the production of food, biofuels and other economically and socially important activities. It might be carried out in part by a neutral, apolitical, "ocean ambassador" who would participate in national and international events to represent and safeguard the interests of the ocean. This might help to galvanise support for marine biotechnology.

Areas for future work

Governments and private-sector organisations around the world have begun to focus on the marine environment and are working to harness its potential, utilising the methodology and knowledge of marine biotechnology. Several features of marine bioresources may nonetheless require new policy work to ensure the translation of new scientific and technological advances into commercially viable products in an environmentally sustainable manner.

The OECD Working Party on Biotechnology has identified three areas of focus for future policy work: governance of marine bioresources and eco-systems, measures and indicators for marine biotechnology, and development of new R&D infrastructure. International collaboration and co-ordination and stakeholder dialogue are expected to feature prominently in this work.

Governance of marine bioresources and ecosystems

Marine biotechnology is predicated on access to marine resources distributed throughout a vast and complex shared ecosystem. An appropriate framework is needed to ensure access to these resources in an environmentally and socially sustainable manner if the potential of marine biotechnology is to be realised.

Measures and indicators for marine biotechnology

Marine biotechnology is poised to make a significant contribution to the bioeconomy. Many countries have included it in national strategies to bring economic and social benefits to their citizens. To measure the contribution of marine biotechnology to the bioeconomy and the return on investment and impact of government policies will require indicators and statistics that are appropriate for marine biotechnology.

Recognising the ecosystem services that marine biotechnology provides (including its non-market value) and the need for the sustainable development of marine bioresources, tools and metrics will also be needed to serve as indicators of "healthy ecosystems" and ensure the fitness of marine bioresources for future generations.

Development of new research and development infrastructure

Marine biotechnology requires a comprehensive understanding of marine bioresources and marine ecosystems. Given the complex nature of the marine environment, it will be necessary to develop new infrastructures for characterising and understanding the potential of ocean bioresources.

Note

1. See Chapter 3, www.sitra.fi/en/natural-resources-strategy, www.pmg.org.za/report/20120222-department-science-technology-grand-challenges-bioeconomy-committee-d, and www.whitehouse.gov/sites/default/files/microsites/ostp/national_bioeconomy_blueprint_april_2012.pdf, accessed August 2012.

ORGANISATION FOR ECONOMIC CO-OPERATION AND DEVELOPMENT

The OECD is a unique forum where governments work together to address the economic, social and environmental challenges of globalisation. The OECD is also at the forefront of efforts to understand and to help governments respond to new developments and concerns, such as corporate governance, the information economy and the challenges of an ageing population. The Organisation provides a setting where governments can compare policy experiences, seek answers to common problems, identify good practice and work to co-ordinate domestic and international policies.

The OECD member countries are: Australia, Austria, Belgium, Canada, Chile, the Czech Republic, Denmark, Estonia, Finland, France, Germany, Greece, Hungary, Iceland, Ireland, Israel, Italy, Japan, Korea, Luxembourg, Mexico, the Netherlands, New Zealand, Norway, Poland, Portugal, the Slovak Republic, Slovenia, Spain, Sweden, Switzerland, Turkey, the United Kingdom and the United States. The European Union takes part in the work of the OECD.

OECD Publishing disseminates widely the results of the Organisation's statistics gathering and research on economic, social and environmental issues, as well as the conventions, guidelines and standards agreed by its members.

OECD PUBLISHING, 2, rue André-Pascal, 75775 PARIS CEDEX 16
(93 2013 01 1 P) ISBN 978-92-64-19423-6 – No. 60501 2013-01